Tiwari, Dubey, Singh, Zayas Saucedo (Eds.)
Luminescence

Also of Interest

Photosynthesis.
Biotechnological Applications with Micro-Algae
Edited by Matthias Rögner, 2021
ISBN 978-3-11-071691-7, e-ISBN 978-3-11-071697-9

Transport of Infrared Atmospheric Radiation.
Fundamentals of the Greenhouse Phenomenon
Boris M. Smirnov, 2020
ISBN 978-3-11-062765-7, e-ISBN 978-3-11-062875-3

Chemical Photocatalysis
Edited by Burkhard König, 2020
ISBN 978-3-11-057654-2, e-ISBN 978-3-11-057676-4

Quantum Electrodynamics of Photosynthesis.
Mathematical Description of Light, Life and Matter
Artur Braun, 2020
ISBN 978-3-11-062692-6, e-ISBN 978-3-11-062994-1

Organophosphorus Chemistry.
Novel Developments
Edited by György Keglevich, 2018
ISBN 978-3-11-053453-5, e-ISBN 978-3-11-053583-9

Luminescence

Theory and Applications of Rare Earth
Activated Phosphors

Edited by
Ratnesh Tiwari, Vikas Dubey, Vijay Singh,
María Elena Zayas Saucedo

DE GRUYTER

Editors
Dr. Ratnesh Tiwari
Dr. C.V. Raman University
Department of Physics, Research and
Development Cell
Kargi Kota
Bilaspur, Chhattisgarh
India
31rati@gmail.com

Dr. Vikas Dubey
Bhilai Institute of Technology
Abhanpur Road, Atal Nagar
Raipur 493661, Chhattisgarh
India
jsvikasdubey.physics@gmail.com

Dr. Vijay Singh
Department of Chemical Engineering
Konkuk University
120 Neungdong-ro
Seoul 05029
Republic of Korea
vijayjiin2006@yahoo.com

María Elena Zayas Saucedo
Department of Physics Research from the
Unison DIFUS
University of Sonora
Blvd. Luis Encinas y Rosales S/N
Col. Centro Hermosillo, Sonora
Mexico
mzayas@difus.uson.mx

ISBN 978-3-11-067641-9
e-ISBN (PDF) 978-3-11-067645-7
e-ISBN (EPUB) 978-3-11-067651-8

Library of Congress Control Number: 2021950695

Bibliographic information published by the Deutsche Nationalbibliothek
The Deutsche Nationalbibliothek lists this publication in the Deutsche Nationalbibliografie;
detailed bibliographic data are available on the Internet at http://dnb.dnb.de.

© 2022 Walter de Gruyter GmbH, Berlin/Boston
Cover image: Myriam Borzee / iStock / Getty Images Plus
Typesetting: Integra Software Services Pvt. Ltd.
Printing and binding: CPI books GmbH, Leck

www.degruyter.com

Preface

Luminescence is an old field of scientific research. This is one research field where diverse application areas exist, which range from radiation monitoring for health and safety, phosphors for lamps and display purposes to X-ray imaging to other means of medical diagnostics. The book Luminescence and its application drawing on the knowledge of an expert team of contributors, provides useful insight into successfully harnessing the properties of luminescence for a range of important applications.

Beginning with an introduction to luminescence, the book goes on to explore the properties, characteristics, synthesis, Recent advances in the development of luminescent lanthanide-based supramolecular material and their applications, Lanthanide based Molecular thermometers, Photoluminescence in semiconductor materials, white light emission phenomenon, Thermoluminescence studies, rare earth activated phosphor etc.

This book provides useful information on various applications of luminescence and various types of luminescence materials.

Contents

Preface —— V

List of Authors —— IX

Ratnesh Das, Sneha Wankar, Imran Khan
Chapter 1
Recent Progresses in the Development of Luminescent Lanthanide-Based Hybrid Entities and Their Applications —— 1

Sivakumar Vaidyanathan, Priyadarshini Pradhan
Chapter 2
Lanthanide-Based Molecular Thermometers: An Overview —— 45

I. V. García-Amaya, L. Castro-Arce, G. A. Limon-Reynosa, J. Manzanares-Martinez, M. C. Acosta-Enríque, Ma. E. Zayas
Chapter 3
Redshift in Tellurite Glasses Doped with Eu^{3+} Due to Heat Treatment —— 109

S. P. Kori, R. Rohatgi, V. Lahariya
Chapter 4
Nanophosphors: Emerging Materials for Forensic Applications —— 131

Kamal K. Kushwah, S. K. Mahobia, Vikas Mishra, Sandeep Chhawra, Ratnesh Tiwari, Neha Dubey, Vikas Dubey
Chapter 5
Luminescence Studies of $Y_2Sr_3B_4O_{12}$ Phosphor Doped with Europium Ion —— 165

L. Castro-A., Ma. E. Zayas-S., Oscar R. Gomez-A., Julio C. Campos, C. Figueroa-N.
Chapter 6
Photoluminescence in Semiconductor Materials —— 171

G. V. S. Subbaroy Sarma, Murthy Chavali, Periasamy Palanisamy, I. B. Shameem Banu
Chapter 7
Synthesis and Luminescent Applications of Rare-Earth-Doped Zinc Nanomaterials —— 181

Vikram Awate, Lokeshwar Patel, Rashmi Sharma, A. K. Beliya, Ratnesh Tiwari, Vikas Dubey, Neha Dubey
Chapter 8
Thermoluminescence Glow Curve Analysis of Mn^{4+}-Doped Barium Yttrium Oxide Phosphor —— 211

Rajani Indrakanti, V. Brahmaji Rao
Chapter 9
Theory of Luminescence and Materials —— 215

Shalini Patil
Chapter 10
Luminescence: Phenomena, Applications and Materials —— 227

Vinod Kumar Verma, K. K. Dubey
Chapter 11
Thermoluminescence and Spectral Studies of Some Geological Crystals —— 233

Index —— 241

List of Authors

Sneha Wankar
Department of Chemistry
Visvesvaraya National Institute of Technology
Nagpur 440010, Maharashtra
India

Imran Khan
Department of Chemistry
Dr. Hari Singh Gour Vishwavidyalaya
Sagar 470003, Madhya Pradesh
India

Ratnesh Das
Department of Chemistry
Dr. Hari Singh Gour Vishwavidyalaya
Sagar 470003, Madhya Pradesh
India
ratneshdas1@gmail.com

Sivakumar Vaidyanathan
Department of Chemistry
National Institute of Technology
Rourkela
Rourkela 769 008
Odisha
India
vsiva@nitrkl.ac.in

Priyadarshini Pradhan
Department of Chemistry
National Institute of Technology
Rourkela
Rourkela 769 008
Odisha
India

I. V. García-Amaya
1Programa de Ingeniería en Geociencias
Universidad Estatal de Sonora
Unidad Hermosillo
Ley Federal del Trabajo S/N
Colonia Apolo
Hermosillo, Sonora C.P. 83100
México

L. Castro-Arce
Departamento de Física
Matemáticas e Ingeniera
Universidad de Sonora
Unidad Sur Navojoa
Sonora
México

G. A. Limon-Reynosa
Departamento de Ciencias Quimico Bilogicas
y Agropecuarias
Universidad de Sonora
Unidad Sur Navojoa
Sonora
México

J. Manzanares-Martinez
Departamento de Investigación en Física
Universidad de Sonora
Blvd. Transversal y Rosales S/N
Colonia Centro
Hermosillo, Sonora, C.P. 83000
México

M. C. Acosta-Enríque
Departamento de Investigación en Física
Universidad de Sonora
Blvd. Transversal y Rosales S/N
Colonia Centro
Hermosillo, Sonora, C.P. 83000
México

Ma. E. Zayas
Departamento de Investigación en Física
Universidad de Sonora
Blvd. Transversal y Rosales S/N
Colonia Centro
Hermosillo, Sonora, C.P. 83000
México

S. P. Kori
Department of Chemistry, Biochemistry and
Forensic Sciences
Amity School of Applied Sciences
Amity University Haryana
Gurugram 122413, Haryana
India

https://doi.org/10.1515/9783110676457-204

List of Authors

R. Rohatgi
Department of Applied Physics
Amity School of Applied Sciences
Amity University Haryana
Gurugram 122413, Haryana
India
rrohatgi@ggn.amity.edu

V. Lahariya
Department of Applied Physics
Amity School of Applied Sciences
Amity University Haryana
Gurugram 122413, Haryana
India
vlahariya@ggn.amity.edu

Kamal K. Kushwah
Department of Applied Physics
Jabalpur Engineering College Jabalpur
Madhya Pradesh
India

S. K. Mahobia
Department of Applied Physics
Jabalpur Engineering College Jabalpur
Madhya Pradesh
India

Vikad Mishra
Department of Physics
Sardar Patel University Balaghat
Madhya Pradesh
India

Sandeep Chhawra
Department of Electronics and
Communication Engineering
MATS University Raipur
India

Ratnesh Tiwari
Department of Physics, Coordinator Research
and Development Cell
Dr. C.V. Raman University
Kargi Kota
Bilaspur, Chhattisgarh
India

Neha Dubey
Department of Physics
Government V.Y.T.PG. Auto. College
Durg 491001, Chhattisgarh
India

Vikas Dubey
Department of Physics
Bhilai Institute of Technology Raipur
Kendri 493661, Chhattisgarh
India
jsvikasdubey@gmail.com

L. Castro-A.
Department of Physics, Mathematics and
Engineering
University of Sonora
Lázaro Cárdenas del Río No. 100
Francisco Villa
Navojoa 85880
Mexico

Ma. E. Zayas-S.
Department of Research in Physics
University of Sonora
Blvd. Transversal y Rosales S/N
Colonia Centro
Hermosillo, Sonora, C.P. 83000
Mexico

Oscar R. Gomez-A.
Department of Health Sciences
Cajeme Campus, Sonora
Mexico

Julio C. Campos
Department of Health Sciences
Cajeme Campus, Sonora
Mexico

C. Figueroa-N.
Department of Industrial Engineering
University of Sonora
Hermosillo Sonora
Mexico

G. V. S. Subbaroy Sarma
Department of Basic Sciences and Humanities
Vignan's Lara Institute of Technology and Science (Affiliated to J.N.T. University, Kakinada)
Guntur 522213, Andhra Pradesh
India

Murthy Chavali
NTRC-MCETRC and Aarshanano Composite Technologies Pvt. Ltd.
Guntur
Andhra Pradesh
India
ChavaliM@gmail.com
ChavaliM@outlook.com

Periasamy Palanisamy
Department of Physics
Nehru Institute of Engineering and Technology
T. M. Palayam
Coimbatore 641105, Tamil Nadu
India

I. B. Shameem Banu
Department of Physics
B. S. A. Crescent Institute of Science and Technology
Vandalur
Chennai 600048, Tamil Nadu
India

Vikram Awate
Department of Physics
Dr. C.V. Raman University
Bilaspur, Chhattisgarh
India

Lokeshwar Patel
Department of Physics,
Atal Bihari Vajpayee University Bilaspur
Bilaspur, Chhattisgarh
India

Rashmi Sharma
Department of Physics
Govt. CV College Dindori
Madhya Pradesh
India

A. K. Beliya
Department of Physics
Govt. CV College Dindori
Madhya Pradesh
India

Ratnesh Tiwari
Department of Physics, Research and Development Cell
Dr. C.V. Raman University
Kargi Kota
Bilaspur, Chhattisgarh
India

Vikas Dubey
Department of Physics
Bhilai Institute of Technology Raipur
Kendri 493661, Chhattisgarh
India
jsvikasdubey@gmail.com

Neha Dubey
Department of Physics
Government V.Y.T.PG. Auto. College
Durg 491001, Chhattisgarh
India

Rajani Indrakanti
Department of Physics
VNRVJIET
Hyderabad, Telangana
India
rajaniindrakanti691@gmail.com

V. Brahmaji Rao
Department of Nanoscience and Technology
School of Biotechnology
DSRF
Hyderabad, Telangana
India
Centre for Nanoscience and Technology
G. V. P. College of Engineering (Autonomous)
Visakhapatnam 530048, Andhra Pradesh
India

Shalini Patil
Department of Physics
Government Autonomous Post Graduate
College
Chhindwara
Madhya Pradesh
India
dr.shalinipatil19@gmail.com

Vinod Kumar Verma
Kendriya Vidyalaya No. 4
Jamnipali
Korba, Chhattisgarh
India

K. K. Dubey
Department of Physics
Government G.B. College Hardibazar
Korba, Chhattisgarh
India

Ratnesh Das, Sneha Wankar, Imran Khan

Chapter 1
Recent Progresses in the Development of Luminescent Lanthanide-Based Hybrid Entities and Their Applications

Abstract: This chapter concentrates predominantly on the recent trends in fabrication and synthesis of luminescent lanthanide hybrids. It also unveils the detailed insight of energy transfer mechanism that proceeds in fundamental lanthanide organic complexes. To achieve the advancement in the basic property of lanthanide complexes in terms of enhanced mechanical strength, better temperature tolerance, desired photophysical properties, synthesized lanthanide complexes are specifically embedded or bonded with organic/inorganic matrix. In this milieu, the synthetic approaches that deal with advance methodologies were taken into consideration. Supplementarily, the application of such advance materials in the real world is profoundly documented in detail.

Keywords: lanthanide ions, TESPIC, ORMOSILs, silica, beta-diketone, mesoporous hybrid

1.1 Introduction

Lanthanide complexes are adeptly beneficial in photonics by virtue of their multifaceted photophysical properties [8] that are markedly relevant to their adoption in fabrication of highly energetic phosphors [47]. They are also projecting themselves as lasers, namely, yttrium aluminum garnet (YAG) [133]. Their relevance in the materials is viable for light amplification, highly efficient optical light-emitting diodes (LEDs), modeling of various electronic displays, imaging of cell organelles, selective sensing of various ions which are important biologically and also environmentally in immunoassays and utilization in medical field [50, 91, 101]. In accordance with IUPAC, elements from cerium to lutetium are entitled as lanthanide series; their electronic configuration is $4f^N$ where $N = 1–14$, and their electronic configuration is featured with +3 oxidation state [82]. The luminescent behavior of lanthanides corresponds to operative transition in terms of $d–f$ transitions (allowed) or reorganization of electrons in the $4f$ shell and appears as $f–f$ transition. The nature of $d–f$ transitions is highly energetic and intense over $f–f$ transition, which is not allowed according to Laporte's selection rule. However, it is allowed under the magnetic dipole–induced transition, and the emission intensity is feeble in nature. The disposition of $4f$ shell is completely shielded away from outer environment by $5s^2$ and $5p^6$ shells. When these lanthanide ions coordinated with ligands containing oxygen or nitrogen, the electronic configuration get perturb. The shielded $4f$ orbitals appear with the unique emission property ranging from visible to near-infrared (NIR) region, longer lifetime of excited state. The position of the emission band of the lanthanide complexes depends on the nature of the ligand and the wavelength of excitation. The resultant luminescence is perceived by means of divergent source of excitation. Thus, few of them listed as excitation by means of electromagnetic radiation show photoluminescence, influenced by electric field resulted in electroluminescence, applied mechanical stress to give rise to triboluminescence, finally chemiluminescence is obtained because of chemical reactions. The concept of crystal field splitting came to an existence with the help of theoretical calculations and atomic quantum theory.

Keywords: lanthanide, photophysical, magnetic dipole, crystal field

Interestingly, furthermore, crystal-field effects are extremely accountable for the ease of selection rule, and thus additional lines broaden some components of the spectral lines. In the first instance, Judd and Ofelt in 1962 [46, 86] streamlined the studies on the intensities of transitions, then secondly, Wybourne well tried to measure the energies up to certain level [20, 137]. These researchers subsequently assigned all electronic sublevels that fall below energy amounted to 40,000 cm^{-1} (250 nm), and the calculated energy levels were molded into very well-known Dieke's diagram. The soundness of lanthanide spectroscopy constitutes bunch of electronic levels brought about by various electronic configurations assigned to $4f^N$ and $4f^N-5d^1$. Taking this into consideration, the noteworthy splitting of crystal field in lower symmetry of electronic levels of trivalent La–Lu is equated to 16,384 for $4f^N$, whereas 180,199 for $f^{N-1}d^1$ configurations. Figure 1.1 demonstrates SLJ electronic levels for previously mentioned electronic levels that markedly considered as higher than 50,000 cm^{-1}. An acronym SLJ implies three quantum numbers where S = ½ multiplied by the number of unpaired present in f-orbital, then L stands for total angular momentum while S is represented as total spin number and can be calculated as S = ½ multiplied by the number of unpaired electrons present in f-orbital. Thus, the total spin–orbit quantum number J can be calculated as [J = from (L + S), (L + S − 1) to (L − S)] [10, 11]. According to Laporte's selection rule, the intraconfigurational transitions are not allowed, and spin rule states that the spin state of electrons must not alter in an optical transition. However, the selection rules are somehow relaxed because of mixing of various wave functions, including vibrational functions. Thus, the resultant f–f transitions are allowed and termed as forced electric dipole transition. In 1953, an eminent worker Dexter [19] has proven that the energy transfer proceeds, which genuinely depends on the distance between donor and acceptor molecules. This work was further expanded by Förster in 1948 [30], referring to the dipole–dipole energy transfer process; apart from this, the three major contributions were identified as follows: electric dipole–dipole transition, electric

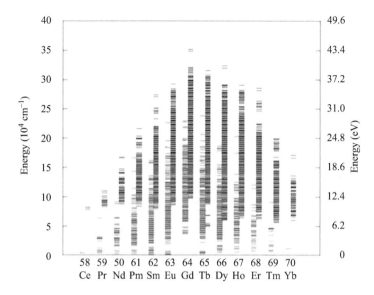

Figure 1.1: The electronic structure of lanthanide ions. Calculated energy levels for all $4f^N$ (left, red) and $4f^{N-1}5d^1$ [right, black] electronic configurations of trivalent lanthanide ions. Reprinted with permission from [10]. Copyright 2017 Wiley.

dipole–quadrupole transition and spin exchange, while electric quadrupole–quadrupole and electric dipole–magnetic dipole have shown negligible significance [100].

Keywords: Dieke's diagram, electric dipole, electric dipole–quadrupole

In this period, another important contribution from eminent researcher Weissman extended the work regarding synthesis of lanthanide complexes [4, 21, 130]. In general, lanthanide complexes are formed when organic ligands get coordinated to the metal ion. As a result, the metal ion gets protected from vibrational coupling and further increases the absorption cross section by means of well-known "antenna effect." The lanthanide metal ions primarily show coordination number 8 or more. Herein the individual emission is as follows: Sm^{3+} emits orange light, Tm^{3+} emits blue light and Dy^{3+} emits white whereas Eu^{3+} emits in red and Tb^{3+} in green region (400–800 nm) apart from this, other listed lanthanide ions are as follows: Pr^{3+}, Nd^{3+}, Sm^{3+}, Dy^{3+}, Ho^{3+}, Er^{3+}, Tm^{3+}, and Yb^{3+} efficiently emit in NIR region (800–1,700 nm). Among them, the Sm^{3+} ion emits typical orange light while Tm^{3+} emits blue light and Dy^{3+} is known for white. In recent years, researches on the subject of NIR emitting lanthanide systems are in demand.

1.2 Detailed Mechanism of the Energy Transfer Process Ensues in the Lanthanide Complexes

The detail process involved in the energy transfer mechanism is very important to discuss, as it lands to very important feature as strength of ligands enacting in complex formation. The indirect way of excitation is an intriguing part, as a ligand grabs energy and transfers its energy to the lanthanide ion. This entire process is termed as "indirect excitation" or "antenna effects." The schematic representation of commencement of energy absorption and to the transfer process is shown in Figure 1.2. Now an organic molecule or ligand in the ground state absorbs requisite amount of energy and then moves from S_0 to S_1 state, followed by transfer of this energy to triplet state of the ligand. This process is known as intersystem crossing. The intramolecular energy transfer takes place from the triplet state of ligand to the emissive/excited state of the lanthanide ion. The entire lanthanide luminescence process is produced by intra-$4f$ transition within ion [49].

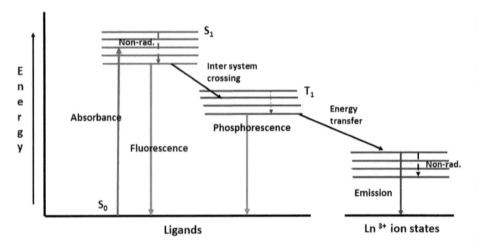

Figure 1.2: Schematic representation of energy transfer process (S, singlet; T, triplet; non-rad., nonradiative transitions).

Within this framework, Dexter proposed the suitability of optimum energy discrepancy among the emissive state of the Ln^{3+} ion and the triplet state energy of the organic ligand is a pivotal factor to facilitate efficient energy transfer [19].

Keywords: antenna effect, energy transfer, emissive

The ideal energy gap between donors and acceptor molecules facilitates energy transfer rate, and in spite of everything when the energy gap is too large, energy transfer is not admissible. On the contrary, the possibility of back energy transfer may occur from the Ln^{3+} ion to the triplet state of the ligand when the reverse condition happens. In general, the energy gap must be conceivable up to 2,000 cm^{-1} to effectively preclude by energy back-transfer [3]. In order to investigate an appropriate energy transfer from the ligands to the Ln^{3+} ion, the absolute energy of the triplet state level of an organic ligand(s) and the convincible energy levels of the Ln^{3+} ion should take place accurately in advance.

Comprehensively, the energy associated with the triplet state of the ligand is an important component to be considered, so experimentally we can compute its value by using phosphorescence spectra of the corresponding gadolinium complex. At excitation, the gadolinium complex emits singlet state fluorescence solely for the ligand, or the triplet phosphorescence, or both, since Gd^{3+} ion do not endow with proper energy levels below 312 nm or 32,000 cm^{-1}, so these levels failed to procure energy from ligand(s). In order to gain a requisite phosphorescence spectrum, the experimental conditions need to be controlled by recording the spectra under inert atmosphere using liquid nitrogen at 77 K [16, 29, 115].

The experimentally recorded phosphorescence spectra are useful for pertaining the triplet state energy of the ligand, and the shortest wavelength of the phosphorescence spectra of Gd^{3+} complex is measured. Simultaneously, the singlet state energy of ligand is estimated from the value of longer wavelength corresponding to absorption spectra of same Gd^{3+} complex. In due course, the entire process of energy transfer from ligand to metal ion would provide an operative way for achieving the high quantum efficiency value, longer lifetime of excited state and sharp emission. The underlying photophysical property marks the lanthanide complexes chiefly valued in sensors and various displays, and enables them to be employed for other applications such as fluorescence microscopy [6, 9]. For synthesizing lanthanide complexes, the commonly used ligands are beta-diketones, aromatic carboxylates, macrocyclic ligands and ancillary ligands (namely, 1,10-phenanthroline, bipyridine and triphenylphosphine oxide).

Keywords: energy gap, ligands, phosphorescence, gadolinium

However, as it is well acquainted for lanthanide complexes due to some shortcomings, these complexes are departed from various practical applications. The probable reason is mainly due to their lesser process stability, low mechanical strength, lesser chemical stability and poor thermal stability [18, 44, 48, 105, 112]. With the intention of solving the persisting drawbacks, some strategies are listed herein to realm the intact luminescent behavior of lanthanide complexes for better rigidity, thermal tolerance and high mechanical strength. Therefore, a detailed study of luminescent lanthanide complex incorporated in different matrixes will be very intriguing part of lanthanide chemistry, and these fundamentals of hybrid materials are expected to show high potential for various applications.

The recent literature in consideration of such synthesized hybrid materials has purported with superior mechanical properties with enhanced ability than the pure molecular lanthanide complex. Moreover, it focuses on the thermal tolerance of hybrid when lanthanide complex embedding in an inorganic/organic matrix cropped up with gratifying thermal stability. The exclusive feature of synthesizing the polyfunctional organic–inorganic hybrid structures through a broadly acclaimed "bottom-up" approach is based on tailoring at the nanosize level of organic (or even bioactive components) and inorganic building blocks [85, 89]. The attainable solution to adopt a synthetic pathway that involves embedding a lanthanide complex in various suitable matrixes is discussed later.

1.3 Development of Hybrids Using Lanthanide Complexes with Ligands Containing Organically Modified Silica (ORMOSILs) as Precursors

The pivotal approach in designing and crafting of such chemically bonded hybrid materials via various functional and bridging molecules is also expected to act reasonably as a chemical linkage explicitly for sensitizing lanthanide ion [7, 17, 33, 53–55, 90, 108, 120, 123, 160].

Keywords: mechanical, thermal, bottom-up

The key point to be noted during drafting such a bridging molecule is the photoactive group, owing to its engagement in hybrid formation. In this domain, a majority of covalently linked group as a bridging molecule is well accomplished as a multifunctional Si–O group, while nitrogen embracing heterocycle chained with carboxylates was reported categorically [52, 129].

The fundamental strategy is to acquire diversity of ligands by implanting a distinctive photoactive ligand onto an especially functionalized silane, a cross-linking reagent. Being relevant to this diversified organic ligands have been functionalized and grafted onto silanes with the aim for achieving desired molecular bridges organically modified silanes (ORMOSILs); the modification paths are easily understood based on different chemical reactions [67, 145, 146, 150]. The heterocyclic compound 2-amino-5(4)-phenylthiazole embodies a reactive hydrogen that reacts with the cross-linking reagent silane 3-(triethoxysilyl) propyl isocyanate (TESPIC) via hydrogen transfer reactions. The resulting silylated linkages (ORMOSILs) are utilized as siloxane network precursors to assemble lanthanide hybrids [35] as shown in Figure 1.3. Crown ethers are compounds comprising cyclic ring with several ether groups, if an oxygen atom is replaced by an amino functional group, then it is termed as aza-crown. Owing to the importance of aza group, some aza-crown ethers comprising 15–18 atoms are successfully reacted with TESPIC with the aim of achieving operating molecular precursors ORMOSILs, some of them are depicted in Figure 1.4 [76, 159].

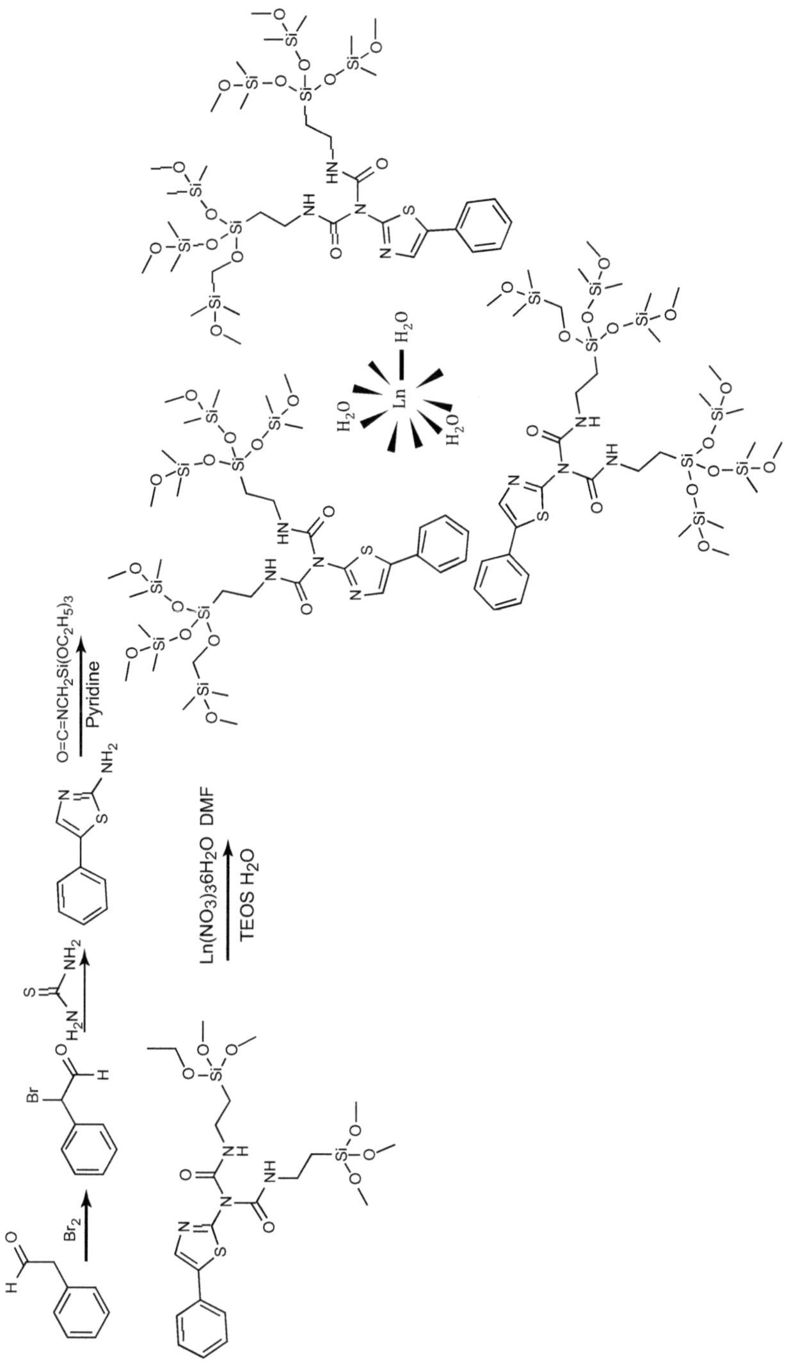

Figure 1.3: Schematic representation for synthesis of 2-amino-5-phenylthiazole and ORMOSILs. Reproduced with permission from [35]. Copyright 2010 Wiley.

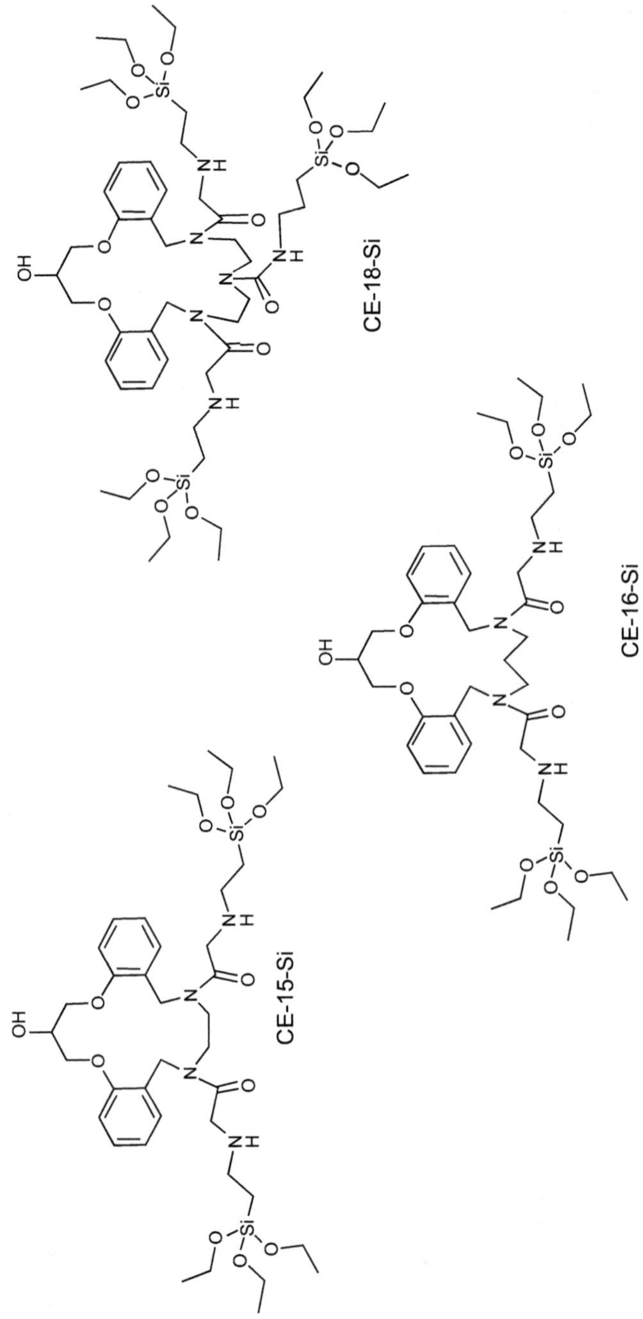

Figure 1.4: All structures showing silylated precursors containing aza-crown. Reproduced from [76], with permission from the Royal Society of Chemistry.

Some hybrids encompassing hydroxyl groups were found to be contributing poor luminescent behavior in terms of quenching of luminescent center, because of increased vibrational energy produced by the hydroxyl groups. In order to overcome this problem, novel macrocyclic compounds like calix[4]arenes were chosen for immobilizing lanthanide complexes over it, their host–guest capability fascinated researchers to dig new entrant in this direction. Thus, it was attempted to alter the phenolic group present on lower side of calix[4]arene using electrophilic reagent TESPIC.

Keywords: ORMOSILs, TESPIC, calix[4]arenes

A successful conversion of active hydroxyl group of *p*-tert-butylcalix[4]arene into urethanesil-grafted bridges was successfully reported as shown in Figure 1.5 [77]. In the same manner, another research group reported the modification of 1,3-bis(2-formylphenoxy)-2-propanol (BFPP) compound via reaction hydrogen transfer nucleophilic addition reaction with TESPIC results in the formation of ORMOSIL compound BFPP-Si as displayed in Figure 1.6. BFPP-Si exerts as a bridging molecule which could possibly coordinate to metal ion (Eu^{3+} or Tb^{3+}) and also expected to form Si–O network with tetraethoxysilane [74].

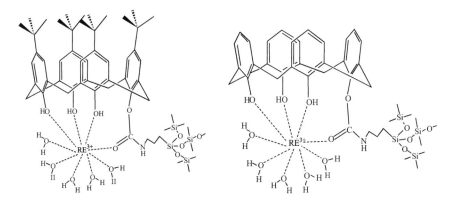

Figure 1.5: The structure of typical calix[4]arene-derived ORMOSILs and their corresponding lanthanide hybrids. Reprinted with permission from [77]. Copyright 2009 American Chemical Society.

The magnification of quantum yield values for lanthanide complexes primarily revolves around the emplacement of triplet state energy of corresponding ligand and its proximity toward emissive state of lanthanide ion [73]. So viewed in this way, another important class of compound as carboxylates also performed well for sensitizing lanthanide ion via antenna effect. The direct modification of carboxylate group is not anticipated as carboxylates very well employed as a bridging ligand. Now aside from carboxylate group keeping unaltered, the amendment of the substituent unit of carboxylate derivatives such as amino, mercapto and hydroxyl is the part of concern [106, 122, 163]. Therefore, many ORMOSILs were derived from various carboxylic acids and lanthanide complexes were embedded in them.

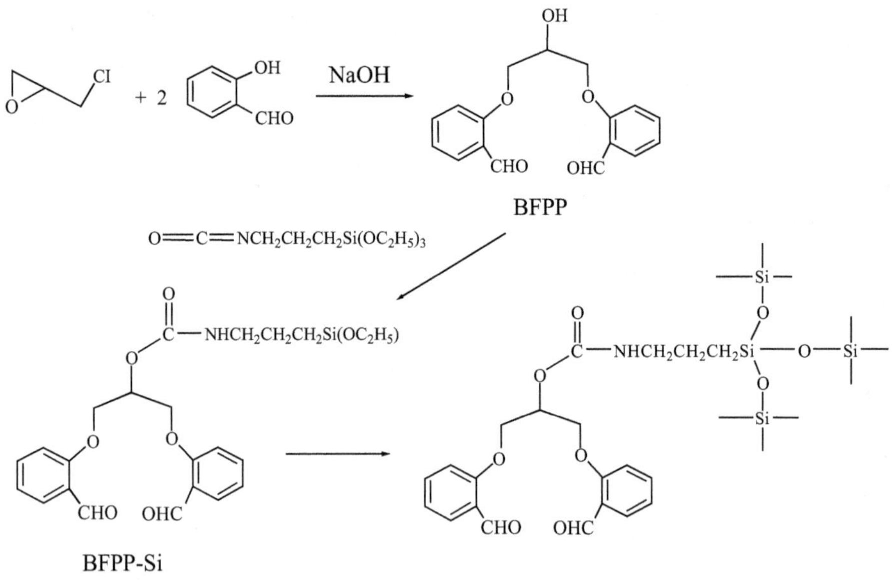

Figure 1.6: Scheme for synthesis of 1,3-bis(2-formylphenoxy]-2-propanol (BFPP) and ORMOSILs [BFPP-Si]. Reprinted with permission from [74]. Copyright 2008 American Chemical Society.

The ligands synthesized for the hybrids are dedicated as ORMOSILs; they are obtained from versatile cross-linking reagents that are comfortably coupled with the organic ligand [147, 153]. The terminal silanol group existing on ORMOSILs reacted with the hydroxyl group actively present on TEOS, and the reaction proceeds via condensation reaction.

In this manner, many research groups are working deliberately in this field for deriving ORMOSILs from pyridine moiety. The amino group located on 2,6-diaminopyridine is a very important entity for selective modification by TESPIC, and then nitrogen containing pyridine group provides a suitable site for three coordinating lanthanide ions [35, 72, 77, 143, 145].

Keywords: BFPP-Si, quantum yield, carboxylates

A hybrid system is composed of bis-silylated-bipy ligand achieved from 4,4-diamino-2,2-bipyridine and three TTA (2-thenoyltrifluoroacetone) ligands. The resulted hybrid entity exhibited longer appreciable lifetime as compared to pure complex.

Another attempt for hybrid formation using ionic liquid is entailed with carboxyl functionalized in the form of luminescent ionogel monolith [145, 155]. It is well acclaimed about the existence of lanthanide complexes generally in the form of polymeric microstructures, and the same was witnessed in the images captured from scanning electron microscopy as shown in Figure 1.7; a branch-like morphology that is prominently visible is derived from carboxylic acid [121, 160].

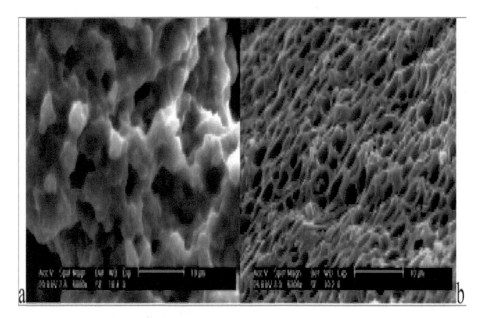

Figure 1.7: Morphologies of different terbium hybrid systems of ORMOSILs obtained with aromatic carboxylate groups ((a) *m*-aminobenzoate; (b) benzoate]. Republished with permission of Royal Society of Chemistry from [120].

The lanthanide beta-diketone complexes are readily formed from lanthanide ions coordinating with diketone. The luminescence obtained for such tris-complexes, where ligands are aliphatic in nature and listed as acetylacetone, trifluoroacetylacetone (TFA), or hexafluoroacetylacetone gives weak luminescence. The possible reason may be due to the large energy gap between the emissive levels of the europium ion and the triplet state energy level of the ligand; the larger energy gap makes unfavorable for the proper energy. Thus, combination of aromatic and aliphatic substituents on the single diketones is reported for intense luminescence.

The aromatic ligands discussed above are a part of the vigorously explored class of compound called beta-diketone; these ligands are reported to be most important sensitizers, especially for europium ion. The combination of various aliphatic and aromatic substitutions on the beta-diketones formed complexes with europium ion exhibited highly intense emission. In all systems, the energy transfer from the ligand beta-diketone to the lanthanide ion was found to be more efficient because of highly favorable triplet state energy placed over the emission state of lanthanide ion. In this locale, lanthanide complex of Eu^{3+}, Nd^{3+}, Sm^{3+} with composition of $Ln(L-OH)_3 \cdot 2H_2O$ (where L = DBM (dibenzoylmethane)) is *o*-hydroxy dibenzoyl methane, covalently grafted over SiO_2 via sol–gel method.

Keywords: bis-silylated-bipy, morphology, beta-diketone

The experimentally obtained quantum yield value in Eu–DBM–Si hybrid over pure complex value enunciates that the energy transfer process that took place in hybrid is more efficient [27]. It is well stated about the infrequent role played by a diurethane; the newly designed hybrid was derived via immobilization of complex, namely Eu(BTFA)$_3$(H$_2$O)$_2$ (BTFA = 4,4,4-trifluoro-1-phenyl-2,4-butanedionate onto poly(oxyethylene)/siloxane diurethane d-Ut (600). An ionic liquid, namely, 1-hexyl-3-methylimidazolium bis (trifluoromethylsulfonyl) imide abbreviated as [C$_6$mim] [Tf$_2$N] coordinated with lanthanide ion well embedded in the pores of a nanoporous silica network results in the formation of a hybrid. One more remarkable list of hybrids is reported, where ionic liquid is [C$_6$mim][Ln(TTA)$_4$] (Ln = Nd^{3+}, Sm^{3+}, Eu^{3+}), while terbium complex is used as [choline]$_3$[Tb(DPA)$_3$] (DPA stands for pyridine-2,6-dicarboxylate) [123].

Other accomplished hybrids were organized using mesoporous silica hybrids via grafting lanthanide complex units in some templates. The identified mesoporous materials are well-ordered/organized and pronounced potential matrices due to the incidence of systematic pore structure, which is responsible for pertinent thermal stability and photostability. In continuation of such interesting hybrid materials, the research extensively carried out using typical PMO stands for periodic mesoporous organosilicas (PMOs) such as MCM-41 and SBA-15/16. Therefore, the functionalized microporous host is well explored for assembling the functional hybrid materials. In detail, some research groups are working deliberately for contributing some significant research work via designing of such hybrid materials using zeolite (ZL) microcrystals [38, 40]. The distinct structure of the ZL tends to support the embedded lanthanide complex, and the structure of ZL is a well-structured booth that appreciably sustenance the lanthanide complex and thus the hybrids are important entrants in numerous practical applications. Not only europium complexes but also synthesized terbium complexes functionalized with silylated form are well encumbered in ZL; they also exhibited useful application as selective sensing of dipicolinic acid. Moreover, the hydroxyl groups present on the quartz plate magnificently coordinated with lanthanide ion [78].

Keywords: hybrid, organosilicas, microporous

A well-prepared nanochannel of reported ZL crystals were reported in literature and shown in Figure 1.8 [13, 127]. The functionalized mesoporous hybrids were in literature owing to their photoactive nature; the designing of such materials started with MCM-41 mesoporous hybrid lanthanide complexes is efficiently incorporated. In continuation of this work, the other researchers were driven with the idea of strong physical and chemical properties to customize films using sol–gel technology [84]. Thus, the MCM-41 mesoporous material covalently bonded with ternary europium turns out to be a promising material [32]. Some other group of researchers worked profoundly and reported MCM-41 mesoporous hybrids with very frequently used ligands for lanthanides as DBM and acetyl acetone ORMOSILs. Thus, the hybrid

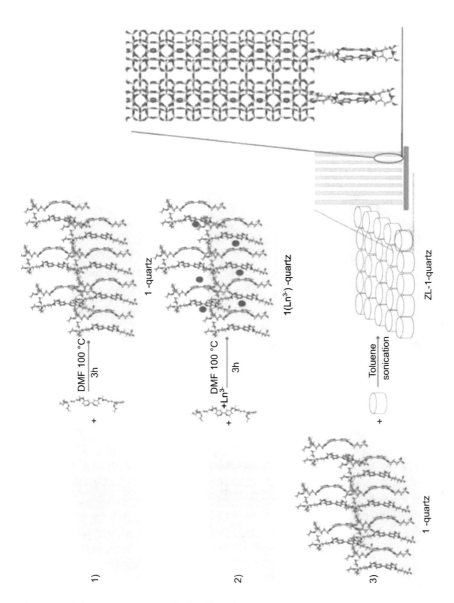

Figure 1.8: Scheme representing the bonding of ORMOSILs to the hydroxyl groups of a quartz substrate and successive formation of oriented open-channel monolayers. Reproduced with permission from [128]. Copyright 2010 Wiley.

materials formed by covalently bonded MCM-41 and lanthanide complexes reveal higher luminescence intensities and longer lifetimes in comparison with pure complexes [145, 154].

An important organic compound, namely, 8-hydroxyquinoline, further functionalized as an organosilane (Q-Si), then conjugated with MCM-41 by means of ligand exchange reaction LnQ3–MCM-41 was finally prepared (Ln = Nd, Yb). These lanthanide hybrid materials exhibited significant NIR emission that again remarks for various applications [26]. Another important result is obtained from complexes of terbium and europium when covalently immobilized over MCM-41. The luminescence intensity and lifetime of these lanthanide hybrids were enhanced as compared to pure complex that further articulates that the triplet state energy of ligand is more favorable for the central metal ion terbium than the europium ion. Such similar type of aromatic carboxylic acid and beta-diketone-type ligands grafted onto MCM-41 has a better place in material chemistry [60, 62]. The importance of ligand Q-Si also reported for its modified form of mesoporous material Q-SBA-15 was synthesized via covalently conjugating with 8-hydroxyquinoline, and the morphology of gained hybrid was found to be of curved cylindrical shape. The emission of these hybrids obtained in the NIR region corresponds to the characteristic emission of Nd^{3+}, Er^{3+}, Yb^{3+}; thus, such highly emissive materials are very useful for strategic designing of optical materials, especially in telecommunication [107].

Keywords: mesoporous, MCM-41, luminescence

Another remarkable hybrid Eu(TTASi-SBA-15)$_3$phen material came out with the characteristic emission of europium at 617 nm corroborated for europium; moreover, high luminescence intensity, enhanced quantum efficiency value and longer lifetime of excited state over pure complex signify that incorporation of 1-10-phen results in the lessening of excessive non-radiative decays by the coordinated water molecules. In the series of several reported beta-diketones, the work was extended by modifying unique blend of organic–inorganic mesoporous luminescent entity by embracing Eu^{3+}/Tb^{3+} covalently attached to the functionalized and well-ordered mesoporous SBA-16 (TTA-S16 and DBM-S16). The resulted hybrid material bpy-Eu-TTA-S16 turned out to have enhanced luminescence intensity with longer lifetime and higher quantum efficiency value [57].

On moving toward the advancement in the hybrid synthesis, the significant role of polymer is an attractive entity as it provides good cross-sectional area for absorption, which simultaneously embraces the lanthanide complex in it. Thus, the use of polymer to enrich the photophysical properties as evident from the enhanced lifetime and quantum efficiency for SBA-15 and SBA-16 hybrid mesoporous is displayed in Figure 1.9 [58].

A PMO is another class brought to being very useful for assimilation of Eu(TTA)$_3$·2H$_2$O into phen-PMO, achieved via the ligand exchange reaction. The result obtained from thermogravimetric analysis of Eu(tta)$_3$phen-S15 hints toward the better thermal

Figure 1.9: High-resolution transmission electron microscopy images of mesoporous hybrid material Eu[TTA-SBA-15]$_3$PMMA. Reprinted with permission from [59]. Copyright 2008 American Chemical Society.

tolerance of hybrid as compared to pure europium complex in a material that possessed better thermal stability than the pure complex [40]. These results also revealed that the chelating organic ligand structure is preserved during the entire surfactant extraction process. Another remarkable method called co-condensation method is highly efficient for synthesized PMO by linking europium or terbium to mesoporous matrix with the aid of modified 4-mercaptobenzoic acid, in company with Pluronic P123 surfactant as an important template [64]. The hybrid material bpy-Ln-Calix-NH$_2$-PMO (Ln^{3+} = Eu^{3+}, Tb^{3+}) was prepared by linking bpy-Ln-Calix-NH$_2$-PMO, where Ln^{3+} = Eu^{3+}, Tb^{3+} to PMOs via Calix-NH$_2$. The resultant hybrid material showed longer lifetime and emission intensity as compared to the pure complex [60].

Keywords: periodic mesoporous organosilicas, Pluronic P123, polymer

In the system Eu-Calix-NH$_2$-PMO, the triplet state energy level gets well or appropriately placed Calix-NH$_2$-PMO at the optimum level, so efficient energy transfer process occurs to sensitize europium and terbium. The enhanced luminescence intensity of the characteristic transition $^5D_0 \rightarrow {}^7F_2$ was observed for europium selectively, including the quantum efficiency value elevated for this material was found to be higher than that of pure bpy-Eu-Calix-NH$_2$ complex as shown in Figure 1.10 [60].

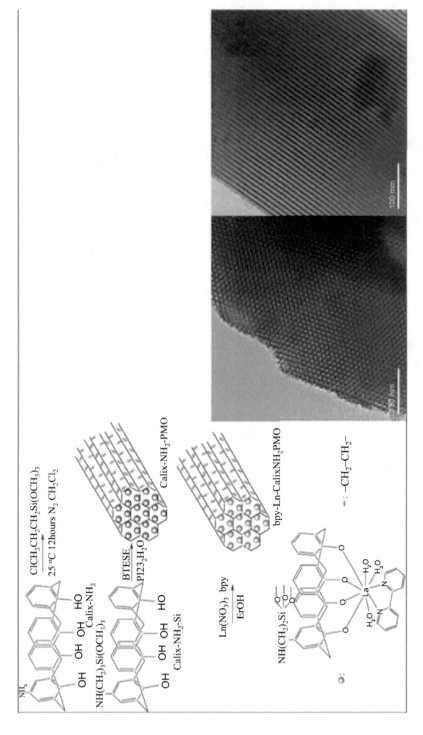

Figure 1.10: Schematic representation of synthesis and HR-TEM images of the ternary mesoporous hybrid bpy-Tb-Calix-NH$_2$-PMO. Republished with permission of Royal Society of Chemistry from [60].

1.4 Development of Hybrids Using Lanthanide Complexes with Ligands Containing Organically Modified Silica (ORMOSILs) onto Polymer Chains

In view of advances in the field of hybrid synthesis, various matrixes such as silica, ZL and MCM-41 are well explored. The assimilation of lanthanide complexes in polymer hosts has been in practice from the last few years [87, 113]. Now incorporation of lanthanide ions/complexes into polymers is also important to discuss. The researchers have already started their important contribution in this field of research. The polymers formed by siloxane particle are covalently bonded. As a result, the nanohybrids were obtained from urea-cross-linked siloxane-POE by the following chemical reactions: aggregation, gelation and aging [69, 83]. The lanthanide hybrids fabricated in polymers with the assistance of solution mixing, blending, polymerization and sol–gel methods [28, 72].

The pure europium complex successfully doped in diurasil network is reported in the literature by the co-condensation method. The amplified photophysical properties well witnessed through important parameters, and the quantitative changes obtained in terms of enhanced emission quantum yield value (up to 36%) and quantum efficiency (57%) as compared to pure complex (29% and 27%, respectively) ensure the relevancy of used polymer.

The encapsulation of lanthanide complex in various matrixes such as silica, organic polymer and biopolymer turns out to be the important entrant that conserved the structure of complex and enhanced photophysical properties. Thus, just diverting the selection of polymer in the form of biocompatible polymer is a significant remark.

Keywords: nanohybrids, sol–gel, diurasil, co-condensation

The microstructures of hybrid materials can be tuned by using different dimensions of silica that are homogenously dispersed in biocompatible polymer chitosan, and the final structure is obtained as Eu^{3+} ion containing chitosan–silica hybrid aerogel. Thus, the reported general emission profile obtained for the core–shell materials further ensures the environments the structure of Eu^{3+} complex is conserved. The future aspect of Ln^{3+}-containing chitosan–silica hybrids is expected in controlled drug delivery or in biological applications [72].

The polymers are again an important entity for producing hybrids the inorganic and organic polymer conjugated through a chemical bond formation at the molecular level. Interestingly, 3-methacryloxypropyltrimethoxysilane (MS) is anticipated as a very good precursor for fabrication of polymer hybrids as MS contains both methacrylate and trimethoxysilyl groups within a molecule. Thus, a series of quaternary- and polymer-based hybrid materials were formed using lanthanide (Eu^{3+}, Tb^{3+}, Dy^{3+}) phenanthroline and precursor as MS and methacrylic acid (MA) [152]. A co-polymer of MS and MA appears to be a strong structural and functional unit for generation of entire chain of organic polymer or inorganic–organic polymer. Further, the coordinates to the lanthanide ion with an ancillary ligand phenanthroline also act as a very efficient sensitizer for the Ln^{3+} ions. The detailed morphological study of the polymer units was carried out, and that finally suggests the formed microstructures to modulate the luminescence properties of hybrid materials. It is important to examine the quaternary phen–Eu^{3+}(Tb^{3+}, Dy^{3+})-MSMA possessing superior photophysical property over pure complex.

Another interesting feature of a polymer spotted is its plastic property, so the copolymer of MMA and maleic anhydrate (MAL) was purposefully brushed up using 3-aminopropyltriethoxysilane (APES). The resultant hybrid phen-Ln-MMA-co-MAL-Si was synthesized by polyhydrolysis followed by condensation between triethoxysilyl group of a modified co-polymer (MMA-co-MAL-APES) and TEOS. The designed and synthesized hybrid exhibited outstanding luminescent behavior with remarkable morphological changes.

Keywords: Eu^{3+}, 3-methacryloxypropyltrimethoxysilane, methacrylic acid

The contribution of terminal ligand 1,10-phenanthroline also reassures the micromorphology of the hybrid system. It demonstrates a well-ordered particle size ranging from 100 to 200 nm. The magnificent image obtained from atomic force microscopy studies shows regular column with a diameter of ~200 nm covalently bonded, and the composites located on rough upper surfaces formed by mutual influences of both polymer chains and silica [155].

Until now, lanthanide hybrids fabricated using various ligands, namely aromatic carboxylic acids, calixarene derivatives, beta-diketonates, ORMOSILs and polymer units have been discussed in detail with the experimental methodology [92, 93, 96, 149]. The obtained distinct morphology for different composition leads to distinguishable photophysical properties. The morphological variability is in terms of homogeneity and arrangements of microstructures. The formation of hybrid as ternary lanthanide embedded in matrixes as inorganic/organic polymer turn out with better photoluminescence properties in terms of strong emission intensities, longer lifetimes and higher luminescence quantum yield as compared to pure complexes, in order to establish the significance of application of polymer chains on microstructures and photoluminescence of hybrids explored. Subsequently, the study reveals that organic polymer chains with various structures are not only coordinated with lanthanide ions by carbonyl group but also generated complete polymeric structure with the inorganic network of $Si-O_2$. The obtained findings further established that all hybrids receive effective intramolecular absorbed energy and thus brilliant characteristic emissions of respective lanthanide ions are obtained. The reaction between the hydroxyl groups on p-tert-butylcalix[4]arene (Calix-Br, Calix AC) and isocyanate group of TEPIC is presented. Thus, polymer chain formed as Calix-Br-Si-Eu (Tb)-PMMA exhibited longest excited state lifetime and highly intense emission with higher value for quantum efficiency for both lanthanides. The coordination of ligand perturbs the electronic distribution in the lanthanide ion with different atomic radii differently, which is also reflected in the co-polycondensation and aging performances during hybrid formation [97]. Another significant pathway is for assembling lanthanide polymer hybrids by the formation of covalent bond between ORMOSILs and different polymers, where indirect interactions or coordinate bonds occur between lanthanide ions and the monomer units of polymer [36, 94, 95, 128]. The chain of organic molecules is linked to inorganic silica through covalently bonded Si–O network; polymers can have a direct impact on the sol–gel process for fabricating the ORMOSIL precursors, which is an important step in the construction of microstructure [156]. A newly reported functional polysilsesquioxane (VPBA-Si) linkage was prepared by the addition reaction (hydrogen transfer) between 4-vinylphenylboronic acid and TESPIC [117, 117, 152]. Furthermore, the hybrid material containing Eu^{3+} and Tb^{3+} was assembled with VPBA-Si and co-polymers using addition polymerization reaction, and then by hydrolysis and polycondensation between the VPBA-Si unit and TEOS.

Keywords: silica, calixarene, photoluminescence

Recently, some research works are especially engrossed on lanthanide hybrids and some other inorganic assemblies were formed as M–O–M, where M = B, Al, Ti. Some research groups have successfully arranged lanthanide hybrids using titania positioned on organic pyridine carboxylic acid linkages, and the same system was cast off to design variety of lanthanide-based hybrids incorporated in silica–alumina or silica–titania composite hosts. In continuation of these, $Eu(Ti-MAB-S15)_2(NTA)_3$ mesoporous matrix associated with Eu $(NTA)_3 \cdot 2H_2O$ is embedded in titania to form a product Ti-MAB-S15 by means of ligand exchange reaction. Therefore, this reported method was instigated to concentrate on the lanthanide-based materials with both ordered mesoporous Si–O networks and unstructured Ti–O networks. The detailed exploration on luminescence characteristics indicate that the formed europium-based titania material $Eu(Ti-MAB-S15)_2(NTA)_3$ with mesoporous nature exhibited enhanced quantum efficiency value and longer lifetime when compared to the complex $Eu(NTA)_3 \cdot 2H_2O$ and titania-based hybrid material Eu(Ti-MAB-S15). A thin film was organized by linking TTA-modified titania precursors to an amphiphilic co-polymer P123 at room temperature [37, 60, 78]. In full swing, polysilsesquioxane bridges can be formed by the modification in TESPIC using ligands such as TTA and TFA (), customarily performed as linkage between Eu^{3+} ions and amalgamated host matrix as Si–O–M (where M = B or Ti).

Keywords: polysilsesquioxane, titania, trifluoroacetylacetone

These xerogels are formed by regulated hydrolysis rates of alkoxyl group situated on TEOS and tributyl borate. The result spectacles that formed lanthanide complexes covalently fixed successfully on the composite inorganic xerogels. These xerogels also exhibited excellent light absorption abilities and very strong emission intensity.

The resulting hybrids such as Si–O–B xerogels showed appreciable photoluminescence properties in terms of high emission, lifetimes and better quantum efficiencies among other xerogels. B and Si were diagonally positioned in the periodic table, and thus these elements show similar features and chemical performances, and it is evident from sol–gel processes. This work was extended to construct other hybrids by introducing various components such as TiO_2, quantum dots (QDs) prepared from ZnS/CdS, ZnSe and GaN, tailored with lanthanide complexes. In this milieu, Eu complex with tunable color appropriately capped with QD ZnSe and, finally, inorganic/organic hybrid nanocrystals (NCs) were successfully synthesized using hot injection method [117, 156, Yan et al., 2011].

The nature of semiconductor is another significant entity for advanced technology, the deep interest in this field because of their unique physical and chemical properties. The important semiconductor ZnS with band gap value of 3.6 eV was reported, which is in line with the requirement as per designing a phosphor or LEDs. In light of an important feature of ZnS, QDs were further compounded with two different cross-linking siloxane 3-mercaptopropyltrimethoxysilane (MS) and TESPIC, then conveniently coordinated to lanthanide ions (Eu^{3+}, Tb^{3+}, Sm^{3+} and Dy^{3+}). The resultant unit TESPIC-MS-SiO_2/ZnS was responsible for getting excited and sensitizing specific lanthanide ion, the characteristic emission observed for this [51, 62]. With advancement in various nanoparticles, their multiple applications in the field of luminescence were explored in detail. Furthermore, the metal sulfide nanoparticles such as CdS, Ag_2S and ZnS were adopted, and they were selectively functionalized using 3-mercaptopropyltrimethoxysilane (MPTMS) to generate MPTMS–CdS, MPTMS–Ag_2S and MPTMS–ZnS composites.

Keywords: xerogels, semiconductor, metal sulfide nanoparticles

The synthesized complex Phen (Bipy)-Eu-TAASi and MPTMS moiety of MPTMS–ZnS/Ag_2S/CdS were assembled by adopting the chemical reaction of co-hydrolysis followed by co-polycondensation. The detailed photophysical and morphological studies reveal that the collusion of nanoparticles generously benefits the entire desired properties as reported and shown in Figure 1.11 [156].

The photosensitive assets of the europium based in inorganic/organic matrix, the hybrid materials thus formed as PMMA-co-Sn_{12} cluster/Eu(TTA)$_3$phen and PMMA-co-Sn_{12} cluster-co-[EuAA(TTA)$_2$phen]. All these materials characterized and found that these are composed of soft chains and are useful for molding into thin films. The nature of the obtained matrix facilitates the formation of transparent thin film onto indium tin oxide (ITO) using the spin coating technique as indicated in Figure 1.12 [22, 23].

The gradual advancement in the field of material science is aided with various multifunctional nanomaterials. The multifunctional mesoporous nanomaterials with distinct magnetic property coupled with enhanced luminescent behavior have shown great potential in the field of biological application as imaging cell lines, as contrast reagent in performing magnetic resonance imaging, drug delivery and many more. So the dual behavior of single material in terms of magnetism and luminescence is

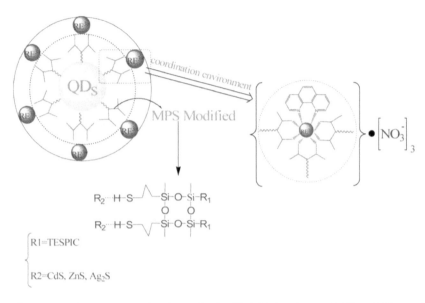

Figure 1.11: Scheme representing the synthesis of hybrid phen-Eu-TAASi-SiO$_2$-MSCdS. Reprinted from [151], with permission from Elsevier.

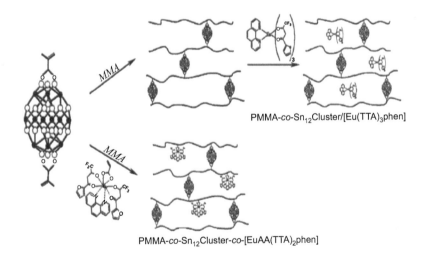

Figure 1.12: Proposed mechanisms for the formation of PMMA-co-Sn$_{12}$ cluster/[Eu(TTA)$_3$phen] and PMMA-co-Sn$_{12}$ cluster-co-[EuAA(TTA)$_2$phen]. Reproduced with permission from [23]. Copyright 2010 Wiley.

achieved by the modified Stöber method followed by layer-by-layer assembly technique. Owing to the magnetic behavior of magnetite Fe$_3$O$_4$, it is potentially active in the field of nanobiotechnology for sensing or imaging and therapeutic interest. Putting all these values in a single framework, hybrid such as Nd(DBM)$_3$phen-MMS (where

MMS = magnetic mesoporous silica) was prepared and employed for its possible application as a laser system or an optical amplifier [25, 27].

The blend of luminescent property of lanthanide complexes and nanomaterial comes out with the unique feature. It is also evident from the material formed by encapsulation of suitably synthesized luminescent europium complexes in multiwalled carbon nanotubes (CNTs).

Keywords: nanomaterials, Stöber method, multiwalled carbon nanotubes, ITO

The appropriate functionalization of exterior wall of CNTs by optimized method leads to the novel class of visible-light-emitting CNT-based lanthanide hybrids and are found to potential entity for developing LED in visible region, or as biodiagnostics as shown in Figure 1.13 [81].

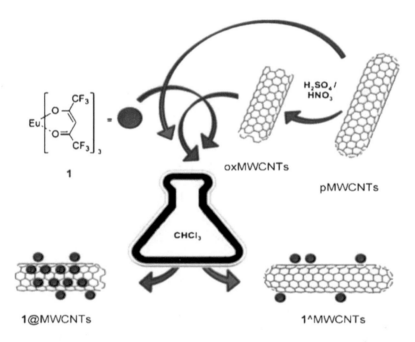

Figure 1.13: Schematic representation of the nano-extraction technique. Reproduced with permission from [81]. Copyright 2011 Wiley.

1.5 An Exploration of Photofunctional Lanthanide Hybrid Materials for Various Applications

The pure intention to design luminescent lanthanide hybrid is to overcome the shortcomings of the pure complex in terms of thermal stability, mechanical strength and photophysical property. In this view, we have discussed herein the various suitable substrates in terms of inorganic or organic adopted selectively for embedding pure lanthanide complexes. The resultant property of hybrid was measured in terms of structural morphology and detailed photophysical properties. It is obvious to discuss the importance of such hybrid material in practical life. The rare-earth-based hybrids displayed pronounced effect because of inimitable luminescent properties; the pertinent research at present is becoming operative and fascinating.

The list of application started with the different photofunctional lanthanide hybrid materials reported as highly luminescent for designing of luminescent solar concentrators (LSC). The novel urea–pyridine-bridged organosilanes were synthesized in the presence of Eu^{3+} salt-bridged silsesquioxanes such as M5–Eu and M6–Eu as shown in Figure 1.14 [34]. The M6–Eu was loaded with 11% w europium with maximum absolute quantum yield value as compared to M5–Eu. The thin film of M6–Eu fabricated by the spin coating technique exhibited one of the utmost reported quantum yields shown by hybrids. The resulted optical conversion efficiency amounted to 4% and thus reported as new potential LSCs. The photograph of another hybrid $F6-Eu^{3+}$ film under excitation at 365 nm displays red light, which can only be observed at film edges through internal reflection used in LSC.

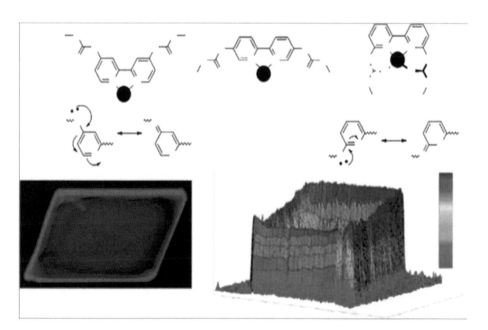

Figure 1.14: (Top) Schematic of plausible interactions between Ln^{3+} ion (full circle) and one diureido-2,2'-bipyridine ligand. (Bottom) Photograph and intensity map of the red pixel of F6–Eu excited at 365 nm. Reprinted with permission from Graffion et al. [34]. Copyright 2011 American Chemical Society.

Keywords: functionalization, lanthanide hybrids, luminescent solar concentrators

The mutual aid of the sol–gel-derived organic/inorganic hybrids is highly luminescent. For such hybrids molded into LSCs, these devices comprise host as a transparent matrix with an active center assimilated in it. When an incidence radiation falls on the matrix, it absorbed light gradually, then emits at particular wavelength, and further transformed by the total internal reflection phenomenon. As shown in Figure 1.15, a part of the emitting signal is vanished at the surface, while only rests at the edge for electrical power generation [15].

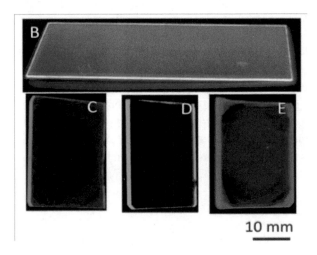

Figure 1.15: Some luminescent solar concentrators (LSCs) under UV irradiation (365 nm) based on (b) a di-ureasil organic–inorganic hybrid doped with [Eu(btfa)$_3$-(MeOH)$_2$ bpta$_2$]$_2$, (btfa = 4,4,4-trifluoro-1-phenyl-1,3-butanedionate, bpta = *trans*-1,2-bis(4-pyridil)ethane and MeOH = methanol) and on bipyridine-based bridged silsesquioxanes (c) lacking metal activator centers and doped with (d) Tb^{3+} and (e) Eu^{3+} ions. Republished with permission of Royal Society of Chemistry from [15].

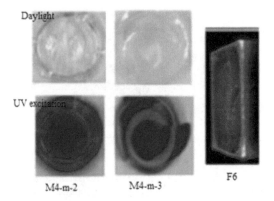

Figure 1.16: M4-m-2, M4-m-3 and F6 and photographs acquired under daylight illumination and UV exposure. Reprinted with permission from [31]. Copyright 2015 American Chemical Society.

In continuation with the development of LSCs or thin films as shown in Figure 1.16, drop cast and spin coating methods are generally used to prepare such films with a q-value of 0.60 ± 0.06; when excited at 345 nm the characteristic emission was obtained. The hybrids were synthesized with the new ethane tetracarboxamide–based organosilane (P4-m) via hydrolysis followed by condensation method, where nucleophilic catalysis carried out in the presence of Eu^{3+} with 2-thenoyltrifluoroacetonate (TTA) led to noteworthy escalation of quantum yield value [31]. Some researchers also reported the methodology in detail; the construction of desired cylindrical LSCs at profitable level was proved to be highly flexible PMMA-based POFs coated with a Eu^{3+}-doped hybrid that stepped forward for the development of lightweighted, strong mechanical strength and highly flexible wave guiding photovoltaics.

Amino clay (AC), magnesium-containing silicate functionalized with amino groups, is recognized as an appropriate inorganic constituent for formulating polychromic luminescent materials. The luminescent lanthanide is embedded in organic or inorganic matrix and its resulting materials by means of assembling sodium 1,2,4,5-benzenetetracarboxylate and Ln^{3+} (Ln = Eu, Tb). Finally, the powdered luminescent hybrids were formed. The obtained gels were easily coated on a round-shaped UV-LED and bright white light achieved as evident from Figure 1.17 [126]. In continuation of this work, luminescent materials were prepared by co-doping deprotonated H_2 Bipydc (2,2′-bipyridine-5,5′-dicarboxylic), AC, and by varying concentrations of Eu^{3+} or Tb^{3+} in composition of AC-bipy-Ln, when excited at 300 exhibited different emission colors [125]. The emission spectra obtained for the hybrids are composed of both line-like emission f–f transition of Eu^{3+}/Tb^{3+} and broadband.

Keywords: organosilane, photovoltaics, amino clay (AC), Ln^{3+}, UV-LED

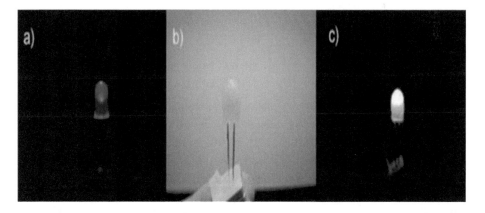

Figure 1.17: Images of (a) commercially available UV-LED cell (λ_{em} = 365–370 nm, 3.2–3.8 V, 20 mA) and (b) UV LED coated with the AC-BTC-Eu1Tb7 hydrogel, a bright white light when the LED is on. Reprinted with permission from [126]. Copyright 2014 American Chemical Society.

A coordinate of (0.30, 0.30) obtained for AC-bipy-Eu1-Tb99 corresponding to "white region" of CIE is ideally close to white light (0.33, 0.33). In modification of state the hydrogel of AC-bipy-Eu1-Tb99 on a commercially available UV-LED cell, bright white light obtained.

Considering the property of the luminescent lanthanide hybrid, another very important application is a luminescent barcoded system. In this view, NIR-emitting lanthanides modified with metal organic frameworks (MOF) were designed, and lanthanide photofunctionalized MIL-100 films exhibited multiple emission bands [134, 165], whose encoding scheme depends on tuning the emission intensities of Ln^{3+} and is considerably diverse from that of the two previously reported MOF barcode systems. A rare-earth ion emitting their characteristic color is given as follows: Eu – red, Tb – green and Sm – orange functionalized with bio-MOF-1 hybrid system by cation exchange. The materials used in barcodes vary with the stoichiometry of Tb^{3+} and Eu^{3+} obtained as Tb_{1-x} Eux@ bio-MOF-1. A barcode was informative based on color code generated with green for Tb^{3+} signal and red for Eu^{3+} signal; their relative intensities reflected in the display to generate discrete barcodes and can be visible with naked eyes [104].

In a biological system, the most common oxidation state of iron is +2 (ferrous)/+3 (ferric). Thus, fabrication of a sensor with selectivity for ferrous over ferric is a difficult task because of its inclination toward oxidation. In this regard, only a fewer fluorescent sensors are reported for selective sensing of Fe^{2+} over Fe^{3+}. The luminescent probe available to detect Fe^{3+} in environmental and biological systems primarily depends on the quenching effect on luminescence produced by MOFs by metal cations. The sensing of Fe^{2+} and even both Fe^{2+} and Fe^{3+} or Fe^{2+} against Fe^{3+} is expected to be explored in detail. An alternative approach to fabricate a highly luminescent nanoscale-sized MOF by encapsulating Eu^{3+} cations in MIL-53-COOH (Al) NCs was developed as an extremely selective and sensitive probe targeting Fe^{3+} cations in aqueous solution.

Keywords: metal organic framework (MOF), lanthanide photofunctionalized, MIL-100

They have examined the potential of designed hybrid named as Eu^{3+}@1 to detect metal cations. The luminescence responses were obtained for various metal cations in aqueous solutions, among them only Fe^{3+} gives noteworthy quenching effect on luminescence in intensity of probe [164]. In continuation of this research, more luminescent hybrids designed as layered-like structure MOF such as MIL-124 or $Ga_2(OH)_4(C_9O_6H_4)$ in which Eu^{3+} ion which is coordinated carbonyl group present in pores, the resultant hybrid system formed as Eu^{3+}@MIL-124 exhibits excellent emission with better fluorescence stability in water and in organic solvent as well. Subsequently, a probe Eu^{3+}@MIL-124 was proven to be a highly sensitive probe for sensing metal ion, especially iron [140]. In order to detect Fe^{2+} and Fe^{3+}, the probe was prepared in the form of paper strips to detect Fe^{2+} and Fe^{3+} in water. For a paper immersed in the aqueous solution of metal ion solution for 1 min and then exposed to air after drying irradiated under UV light, the emitted colors gradually change from faint red to dark red, and finally black in the presence of Fe^{3+}.

The hybrid Tb^{3+}@Cd-MOF was employed for sensing metal cations by examining their resulted luminescent intensity in the presence of Fe^{3+} cation. It is worthy to note that Fe^{3+} can quench the fluorescence intensity extraordinarily [131]. Simultaneously, another diketone-based sensitive hybrid [DBM-Yb-ZA]-NTA-Si-Eu was examined for selective sensing of cation. It is observed that the introduction of Ag^+, K^+, Pb^{2+}, Al^{3+}, Na^+, Mg^{2+}, Zn^{2+} Cd^{2+}, and Hg^{2+} slightly changes the luminescent intensity of [DBM-Yb-ZA]-NTA-Si-Eu, while Co^{2+}, Fe^{2+}, Ni^{2+} and Cu^{2+} have shown different quenching effects on the luminescent intensity of probe. It is interesting to conclude that the demonstrated Fe^{3+} quenches the luminescent intensity of bio-MOF-1 remarkably [109].

The sensing of various ions is with reference to environmental and biological importance. In this context, the element copper is an essential component for biological processes. The optimum concentration of copper in human body is very important, as the deficiency of copper causes disease like myocardial infarction, osteoporosis and vitiligo while excess concentration of copper would cause disorder such as Alzheimer's disease, Wilson's disease and Menkes disease.

Keywords: MIL-124, Eu^{3+}@MIL-124, Tb^{3+}@Cd-MOF, luminescent intensity, sensing

Therefore, the selective sensing of copper in the biological system is an important field of research; in this view, very useful Eu^{3+}@bio-MOF-1 was designed for selective sensing of copper ion in the biological system in the presence of various other ions. The selectivity of Eu^{3+}@bio-MOF-1 hybrid system was employed for sensing different metal ions in DMF solution. The existence of Cu^{2+} quenched the emission of FAM, while the luminescent intensity of Eu^{3+} enhanced [132]. The polyurethane foams (PUFs) are useful as sorbent for the separation of various metals. The structural feature of PUF can be observed for high porosity, worthy mechanical strength and enable lanthanide to accommodate in it. Thus, introduction of $Eu(TTA)_3 \cdot L$ tetramethoxysilane and then finally assembling all components altogether to form a hybrid Eu–PUF. It is highly sensitive and is considered as the coordination between Cu^{2+} and EDTA, which resulted in the "off–on" process will recyclable the property of probe [167].

Another matrix is named as Uio-66(Zr)-$(COOH)_2$, where encapsulation of Eu^{3+} into the pores in a ligand molecule is availed with free carboxyl group and is acquiescent to coordinate with metal cations. The hybrid as [Eu^{3+}@Uio-66(Zr)-$(COOH)_2$] was established as a fluorescent probe and selective sensor for Cd^{2+}, displaying high selectivity, sensitivity and fast detection time as presented in Figure 1.18 [42]. Furthermore, when a hybrid was irradiated under UV light in the presence of analyte, only Cd^{2+} ion can prompt a red-colored luminescence, while no visible alteration was witnessed on addition of other metal ions. The probable mechanism of sensing can be exemplified as Cd^{2+} ions may interact with the Lewis base oxygen of the carboxylic in hybrid system, which facilitates the energy transfer from the located ligands to Eu^{3+} ion. The facilitated energy transfer from ligand to Eu aided the emission of metal ion. Cd^{2+} ion also has a heavy atom effect to endorse the intersystem crossing process [42].

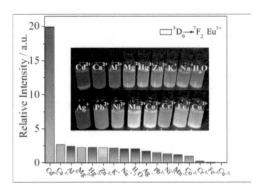

Figure 1.18: The relative intensities of $^5D_0 \rightarrow {}^7F_2$ at 614 nm for $Eu^{3+}@1$ dispersed in various metal ion aqueous solutions (10 mM) when excited at 322 nm. Republished with permission of Royal Society of Chemistry from [42].

Keywords: polyurethane foams (PUFs), Uio-66(Zr)-(COOH)$_2$, fluorescent probe

An encapsulation of carbon dots (CDs) in MOF-253 resulted as CDs@MOF-253 turned out with excellent optical properties furthermore assimilated Eu^{3+} in CDs@MOF-253, the fluorescent sensor Eu/CDs@MOF-253 engaged for ratiometric and colorimetric to detect Hg^{2+} with high sensitivity and selectivity. The probe Eu^{3+}/ CDs @ MOF-253 spectacles as adsorbent for Hg^{2+} [140, 140]. Interestingly, a recyclable probe microsphere SiO$_2$@EuTTA@ZIF-8 was examined for potential sensing of metal ions. The probe SiO$_2$@EuTTA@ZIF-8 in the form of microspheres was reused several times after simply washing the chemosensors with fresh MeCN after the sensing process as shown in Figure 1.19 [70].

The MOF-like zeolitic imidazolate frameworks are topologically isomorphic with ZL. Lanthanide complexes (Eu-TTA) and MOFs (ZIF-8) to cover the unit of silica fabricated the heterostructure sensor. It can be observed from Figure 1.20 that only the presence of Cu^{2+} quenched the emission intensity of probe for the selective sensing and recognition of Cu^{2+} in aqueous media [71]. The luminescent enrichment of Ag^+ on Eu^{3+} @MIL-121 was exploited for sensing Ag^+ ion as shown in Figure 1.21. The results indicate that Eu^{3+}@MIL-121 can selectively sense Ag^+ ions through fluorescence enhancement; this observation is quite rare in luminescent MOFs [41]. Fluoride is considered as a serious health hazard in the biological system, which may lead to some diseases in humans; however, fluorine is broadly adopted in organic synthesis. The consistence efforts were taken to design a simple and cost-effective sensor to detect F^-, for instance, solid-state ion-selective electrodes or devices. In this manner, the progress and groundwork on a

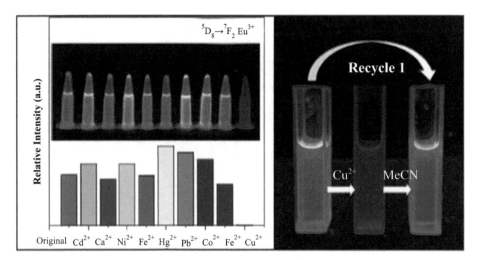

Figure 1.19: (Left) Relative intensities of transition $^5D_0-^7F_2$ at 613 nm for SiO$_2$@EuTTA@ZIF-8 when dispersed in various metal ion aqueous solutions. (Right) Photograph of SiO$_2$@EuTTA@ZIF-8 microspheres recycled for Cu^{2+} sensing. Republished with permission of Royal Society of Chemistry, from [70].

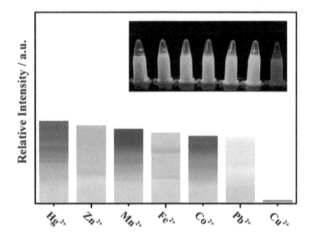

Figure 1.20: The relative intensities of MIL-53-L dispersed in various metal ion aqueous solutions and excited at 337 nm. *Inset*: The photographs under UV light irradiation. Reprinted from [71], with permission from Elsevier.

fluorescent sensor plays a vital role in selective recognition and of producing signals has received enormous attention.

A new polydentate-type ligand derived from *N,N'*-bis(4,4-diethoxy-9-oxo-3-oxa-8,10-diaza-4-siladodecan-12-yl)pyridine-2,6-dicarboxamide (L) was assembled to form a novel hybrid material containing lanthanide complex. The hybrid is designated

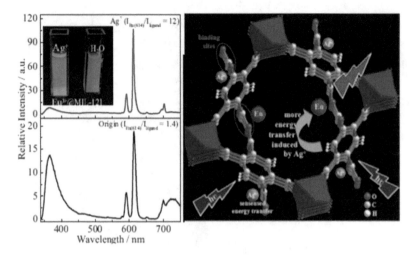

Figure 1.21: The excitation spectra of Eu^{3+}@MIL-121 in the absence (*black*) and presence of Ag^+ (*red*) in aqueous solution. Republished with permission of Royal Society of Chemistry, from [68].

as follows: ASNs-Eu/ASNs-Tb and MSNs-Eu/MSNs-Tb containing amide groups are well coordinated to Eu^{3+}/Tb^{3+} resolutely to sensitize. Both these nanoprobes are observed to fine probe for selective sensing of fluoride ions.

Keywords: carbon dots (CDs), CDs@MOF-253, sensor, luminescent MOFs, fluoride

The noteworthy properties obtained with higher thermostability with lower detection limit are specific for the fluoride ion [168]. The hybrids showed extremely sensitive fluorescence response toward F^- to quench the luminescence of Tb^{3+} via hydrogen bond formation between the ligand and F^-.

Apart from discussing the usefulness and adverse effect of fluoride ions, another compound sodium hypochlorite (NaClO) exhibited its widespread applications in our lives for water disinfection. Nevertheless, the residual part of NaClO as ClO^- in excess quantity is very harmful to the health. The studies have presented that lanthanide complexes can be effectively located onto single-walled nanotube (SWNT) surfaces covalently. The resultant SWNT-based lanthanide complexes come out with advanced properties compared to traditional lanthanide complexes. Newly fabricated SWNT-based lanthanide luminescent hybrids were employed for sensing of ClO^- selectively in the presence of various anions in water. The selective sensing can be evident from quenching of characteristic emission of europium ion as shown in Figure 1.22 [119].

A simple, rapid selective strategy for the ratiometric sensing of carbonate ions is based on the porous material Eu/Pt-MOFs [110]. The mechanism of sensing CO_3^{2-} ion can be perceived by the interaction of CO_3^{2-} with probe results in abruptly enhanced luminescence intensity of europium ion. It can also quantify from the increased value of intensity ratio. Another remarkable hydrogels were reported by

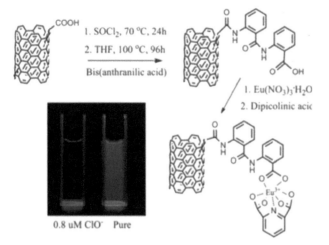

Figure 1.22: Synthetic scheme for the single-walled carbon nanotube covalently modified with a luminescent europium complex [SWNT-4]. Republished with permission of Royal Society of Chemistry, from [119].

doping lanthanide complexes into a polymer hydrogel poly(2-acrylamido-2-methyl-1-propanesulfonicacid). They exhibited very fascinating and operative self-healing enactment in the absence if any external stimulus and reversible "on–off" luminescence intensity interchanging elicited by an exposure to acid–base vapors, whose key observation proceeds with the protonation of organic ligands that compete with full coordination to Ln^{3+} ion [162].

As a similarity of luminescent molecular organic framework, MOFs perform as host–guest materials and highly available for entrapping luminescent lanthanide complexes and form LnMOFs. The pores are actually a kind of within microsized porous ZL when these pores entrapped with LnMOFs are used for detection of toxic gases and vapors of volatile organic compounds (VOCs).

Keywords: single-walled nanotubes (SWNT), LnMOFs, volatile organic compounds (VOCs)

Generally, the designing of such a sensor material for the detection of VOCs is difficult to paradigm and thus very less work was carried out. A simple and robust framework of Eu^{3+}-beta-diketonate complex coupled with nanozeolite L was highly active for the detection of basic molecule in vapors as reported in the literature [56]. In this sight, another important nanohybrid Eu^{3+}@ UiO-bpy was also reported as an efficient sensor for VOCs with alike chemical structures and physical properties (Figure 1.23). The emission profile of Eu^{3+}@ UiO-bpy after the accommodation of different VOCs shows sensitive ratiometric luminescence responses. The resultant comparable ratiometric values were explained in terms of intensity ratios [166].

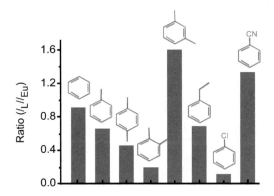

Figure 1.23: Eu^{3+}@UiO-bpy after the encapsulation of different aromatic VOCs. The emission spectra [λ_{ex} = 355 nm] are normalized to the intensity of the ligand emission. Republished with permission of Royal Society of Chemistry, from [166].

The electrochemiluminescence (ECL) is another remarkable luminescence phenomenon induced by electrochemical species as luminophore generated at electrodes, which further undergoes high-energy electron transfer reaction in order to form light-emitting excited states. The electrochemically generated intermediates undergo a highly exergonic reaction in order to produce a high electronically excited state which emits radiation while species reaches the lower state of energy. The ECL phenomenon was generally produced via homogeneous electron transfer or redox reactions; both ions are produced on respective electrodes or at microelectrode arrays [2, 5, Bertoncello et al., 2009, 79, 98, 114]. The mechanistic reactions that occur are represented as follows:

$$\text{Luminophore(L)} \rightarrow \text{luminophore(L}^+\text{)} + e^- \tag{1.1}$$

$$\text{Luminophore(L)} + e^- \rightarrow \text{luminophore(L}^-\text{)} \tag{1.2}$$

$$\text{Luminophore(L}^+\text{)} + \text{luminophore(L}^-\text{)} \rightarrow \text{excited luminophore*(L*)} \tag{1.3}$$

$$\text{Excited luminophore*(L*)} \rightarrow \text{luminophore(L)} + h\nu \text{ (light emitted)} \tag{1.4}$$

It can be witnessed from the ECL reactions that how electrical energy is gradually converted into radiative energy. The reactive intermediates formed from stable precursors are present at the vicinity of electrode surface.

Keywords: nanozeolite L, Eu^{3+}@ UiO-bpy, electrochemiluminescence (ECL), luminophore, electrode

These intermediates undergo reactions under variety of appropriate conditions to form excited state species that subsequently emit light while coming to the ground state. The ECL was generated using two different luminophore precursors, for example,

A and D. The ECL obtained by the cross-reaction of radical cation of one species (A) and radical anion of second species (D) is as follows:

$$A + e^- \rightarrow A^{-*} \text{ (reduction at electrode)} \quad (1.5)$$

$$D - e^- \rightarrow D^{+*} \text{ (oxidation at electrode)} \quad (1.6)$$

$$A^{-*} + D^{-*} \rightarrow A^* + D \text{ (A or D) (excited state formation)} \quad (1.7)$$

$$A^* \text{ (or } D^*) \rightarrow A \text{(or D)} + h\nu \text{ (light emission)} \quad (1.8)$$

The ECL quantum efficiency (ϕ_{ecl}) is an important parameter to estimate the amount of luminophore species generated and results in luminescence. The ϕ_{ecl} is defined as the ratio of the number of photons emitted to the number of annihilation reactions that occur between the oxidized and reduced species [99, 135]:

$$\phi_{ecl} = \frac{\int_t^0 I \, dt}{\int_t^0 i_{c,a} \, dt} = \frac{\int_t^0 I \, dt}{Q_{c,a}} \quad (1.9)$$

where I stands for the total ECL integrated intensity over a finite period of time, t represents photons per sec. $i_{c,a}$ is defined as the integration of the cathodic and anodic current as (i_c) and (i_a), respectively, over the same period of time that results in the total cathodic and anodic charges ($Q_{c,a}$).

The electrochemically generated chemiluminescence involves the production of light at or at the vicinity of electrode surface by electrochemically generating species that are capable of generating excited states [136, 161]. The ECL resonance has been widely used in fabrication of electrochemical sensors, biosensors, immunosensors and arrays for variable inorganic, organic and analysis of biomolecules. The trivalent lanthanides display high efficiency, large shifts between absorption and emission maxima (like 300 nm for Eu^{3+}) and narrow emission spectra.

Keywords: quantum efficiency, electrochemical sensors, biosensors, immunosensors

In this ambience, the interesting phenomenon ECL exhibits rare-earth elements like europium (Eu) when their chelates, cryptates and mixed-ligand chelate/cryptate complexes were studied using persulfate and other precursors as a "reductive–oxidative" coreactant [43, 45, 45, 80, 102, 118, 138].

A very efficient electroluminescent material such as polymer LEDs was prepared with the following architecture: ITO/PEDOT-PSS/PPP-OR11/Ca/Al 40:50:5:200 [43]. The ECL method shows versatile advantages in terms of good temporal and spatial control, wide dynamic range, high selectivity and sensitivity, low background noise, low consumption unselective photoexcitation, scattered light without interference and economic instrumentation with fast sample analysis [12].

Keywords: electroluminescent, polymer light-emitting diodes (PLEDs)

1.6 Conclusion

In this research, we have documented the importance of organic/inorganic lanthanide hybrids with a special focus on their synthetic strategies such as sol–gel processing, grafting of synthesized lanthanide complexes onto the inner networks of mesoporous materials, units on polymer chains, non-silica or mixed matrices. The special emphasis on the photofunctional properties of lanthanide-based hybrid materials was conferred. The synthesized hybrid adopting sol–gel methodology and/or other processes was documented well with their application in LSCs, selective sensing of cations/anions with special reference to biological and environmental importance. In this chapter, we are limited to very few applications, while we are waiting for important contribution from such materials in future.

References

[1] Am, J. (1972). Chem. Soc., 94(18), 6324–6330.
[2] Bard, A. J., Faulkner, L. R. (2004). Electrochemical Methods Fundamentals and Application. 2nd, Wiley.
[3] Beeby, A., Faulkner, S., Parker, D., Williams, J. A. G. (2001). Sensitised luminescence from phenanthridine appended lanthanide complexes: analysis of triplet mediated energy transfer processes in terbium, europium and neodymium complexes, J. Chem. Soc., Perkin Trans., 2, 1268–1273.
[4] Bekiari, V., Lianos, P. (1998). Strongly Luminescent Poly(ethylene glycol)-2,2′-bipyridine Lanthanide Ion Complexes. Adv. Mater., 10(17), 1455–1458.
[5] Bezman, R., Faulkner, L. R., Mechanisms of chemiluminescent electron-transfer reactions. V. Absolute measurements of rubrene luminescence in benzonitrile and N,N-dimethylformamide
[6] Binneman, K. S. (2005). Rare-Earth Beta-Diketonates, Handbook on the Physics and Chemistry of Rare Earths. Gschneidner Jr K. A., Bunzli J. C. G., Pecharsky V. K. Elsevier, Amsterdam, 35, ch. 225, 107–272.
[7] Binnemans, K., Lenaerts, P., Driesen, K., Gorller-Walrand, C. (2004). A luminescent tris (2-thenoyltrifluoroacetonato)europium(iii) complex covalently linked to a 1,10-phenanthroline-functionalised sol–gel glass. J. Mater. Chem., 14, 191–195.
[8] Blasse, G., Grabmaier, B. C. (1994). Luminescent Materials. Springer-Verlag Berlin Heidelberg, Berlin, Germany, 33–70.
[9] Bunzli, J. C. G. (2010). Lanthanide luminescence for biomedical analyses and imaging. Chem. Rev., 110(5), 2729–2755.
[10] Bunzli, J. C. G. (2017). Rising stars in science and technology: luminescent lanthanide materials. Eur. J. Inorg. Chem., 2017(44), 5058–5063.
[11] Bünzli, J. C. G. (2019). Lanthanide photonics: shaping the nanoworld. Trends Chem., 1, 751–762.
[12] Bünzli, J. C. G., Piguet., C. (2005). Taking advantage of luminescent lanthanide ions. Chem. Soc. Rev., 34, 1048–1077.
[13] Cao, P. P., Wang, Y. G., Li, H. R., Yu, X. Y. (2011). Transparent, luminescent, and highly organized monolayers of zeolite. L J. Mater. Chem., 21, 2709–2714.

[14] Carlos, L. D., Ferreira, R. A. S., De Zea Bermudez, V., Ribeiro, S. J. L. (2009). Lanthanide-containing light-emitting organic–inorganic hybrids: a bet on the future. Adv. Mater., 21(5), 509–534.

[15] Correia, S. F. H., Bermudez, V. D. Z., Ribeiro, S. J. L., Andre, P. S., Ferreira, R. A. S., Carlos, L. D. (2014). Luminescent solar concentrators: challenges for lanthanide-based organic–inorganic hybrid materials. J. Mater. Chem. A, 2, 5580–5596.

[16] Crosby, G. A., Whan, R. E., Alire, R. M. (1961). Intramolecular energy transfer in rare earth chelates. Role of the triplet state. J. Chem. Phys., 34, 743.

[17] Dahmouche, K., Carlos, L. D., Santilli, C. V., Pulcinelli, S. H., Craievich, A. F. (1999). Small-angle x-ray scattering study of sol–gel-derived siloxane–PEG and Siloxane–PPG hybrid materials. J. Phys. Chem. B, 103(24), 4937–4942.

[18] De Bettencourt-Dias, A. (2007). Lanthanide-based emitting materials in light-emitting diodes. Dalton Trans., 2229–2241.

[19] Dexter, D. L. (1953). A theory of sensitized luminescence in solids. J. Chem. Phys., 21, 837.

[20] Dieke, G. H. (1968). Spectra and Energy Levels of Rare Earth Ions in Crystals, Crystals. Interscience Publishers, New York.

[21] Driesen, K., Deun, R. V., Gorller-Walrand, C., Binnemans, K. (2004). Near-infrared luminescence of lanthanide calcein and lanthanide dipicolinate complexes doped into a silica–PEG Hybrid Material. Chem. Mater., 16(8), 1531–1535.

[22] Fan, W. Q., Feng, J., Song, S. Y., Lei, Y. Q., Zheng, G. L., Zhang, H. J. (2010). Synthesis and optical properties of europium-complex-doped inorganic/organic hybrid materials built from oxo–hydroxo organotin nano building blocks. Chem.–Eur. J., 16(6), 1903–1910.

[23] Fan, W. Q., Feng, J., Song, S. Y., Lei, Y. Q., Zhou, L., Zheng, G. L., Dang, S., Wang, S., Zhang, H. J. (2010). Near-infrared luminescent copolymerized hybrid materials built from tin nanoclusters and PMMA. Nanoscale, 2, 2096–2103.

[24] Feng, J., Fan, W. Q., Song, S. Y., Yu, Y. N., Deng, R. P., Zhang, H. J. (2010). Fabrication and characterization of magnetic mesoporous silica nanospheres covalently bonded with europium complex. Dalton Trans., 39, 5166–5171.

[25] Feng, J., Song, S. Y., Deng, R. P., Fan, W. Q., Zhang, H. J. (2010). Novel multifunctional nanocomposites: magnetic mesoporous silica nanospheres covalently bonded with near-infrared luminescent lanthanide complexes. Langmuir, 26(5), 3596–3600.

[26] Feng, J., Song, S. Y., Fan, W. Q., Sun, L. N., Guo, X. M., Peng, C. Y., Yu, J. B. Y. N., Zhang, H. (2009). J. Micropor. Mesopor. Mater., 117(1–2), 278–284.

[27] Feng, Y., Li, H. R., Gan, Q. Y., Wang, Y. G., Liu, B. Y., Zhang, H. J. (2010). A transparent and luminescent ionogel based on organosilica and ionic liquid coordinating to Eu^{3+} ions. J. Mater. Chem., 20, 972–975.

[28] Fernandes, M., Bermudez, V. D., Ferreira, R. A. S., Carlos, L. D., Martins, N. V. (2008). Incorporation of $Eu(TTA)_3(H_2O)$ complex into a co-condensed d-u(600) /d-u(900) matrix. J. Lumin., 128(2), 205–212.

[29] Filipescu, N., Sager, W. F., Serafin, F. A. (1964). Substituent effects on intramolecular energy transfer. II. Fluorescence spectra of europium and terbium β-diketone chelates. J. Phys. Chem., 68, 3324–3346.

[30] Förster, T. (1948). Intermolecular energy migration and fluorescence. Ann. Phys., 437, 55–75.

[31] Freitas, V. T., Cojocariu, L. S. F., A. m., Cattoën, X., Bartlett, J. R., Le Parc, R., Bantignies, J., Chi Man, M. W., André, P. S., Ferreira, R. A. S., Carlos, L. D. (2015). Eu^{3+}-Based Bridged Silsesquioxanes for Transparent Luminescent Solar Concentrators. ACS Appl. Mater. Interf., 7(16), 8770–8778.

[32] Gago, S., Fernandes, J. A., Rainho, J. P., Ferreira, R. A. S., Pillinger, M., Valente, A. A., Santos, T. M., Carlos, L. D., Ribeiro-Claro, P. J. A., Goncalves, I. S. (2005). Highly Luminescent Tris(β-

diketonate)europium(III) Complexes Immobilized in a Functionalized Mesoporous Silica. Chem. Mater., 17(20), 5077–5084.

[33] Goncalves, M. C., Silva, N. J. O., Bermudez, V. D., Ferreira, R. A. S., Carlos, L. D., Dahmouche, K., Santilli, C. V., Ostrovskii, D., Vilela, I. C. C., Craievich, A. F. (2005). Local structure and near-infrared emission features of neodymium-based amine functionalized organic/inorganic hybrids. J. Phys. Chem. B, 109(43), 20093–20104.

[34] Graffion, J., Cattoen, X., Chi Man, M. W., Fernandes, V. R., Andre, P. S., Ferreira, R. A. S., Carlos, L. D. (2011). Modulating the photoluminescence of bridged silsesquioxanes incorporating Eu^{3+}-Complexed n,n′-Diureido-2,2′-bipyridine isomers: application for luminescent solar concentrators. Chem. Mater., 23(21), 4773–4782.

[35] Guo, L., Yan, B. (2010). Chemical-bonding assembly, physical characterization, and photophysical properties of lanthanide hybrids from a functional thiazole bridge. Eur. J. Inorg. Chem., 2010(8), 1267–1274.

[36] Guo, L., Yan, B. (2011). Photoluminescent rare earth inorganic–organic hybrid systems with different metallic alkoxide components through 2-pyrazinecarboxylate linkage. J. Photochem. Photobiol., A, 224(1), 141–146, (b) Li Y. J. and Yan B., Preparation, characterization and luminescence properties of ternary europium complexes covalently bonded to titania and mesoporous SBA-15, J. Mater. Chem., 2011, 21, 8129–8136.

[37] Guo, M., Yan, B., Guo, L., Qiao, X. F. (2011). Cooperative sol–gel assembly, characterization and photoluminescence of rare earth hybrids with novel dihydroxyl linkages and 1,10-phenanthroline. Colloids Surf., A, 380(1–3), 53–59.

[38] Guo, X. M., Guo, H. D., Fu, L. S., Carlos, L. D., Ferreira, R. A. S., Sun, L. N., Deng, R. P., Zhang, H. J. (2009). Novel near-infrared luminescent hybrid materials covalently linking with lanthanide [Nd(III), Er(III), Yb(III), and Sm(III)] Complexes via a Primary β-Diketone Ligand: Synthesis and Photophysical Studies. J. Phys. Chem. C, 113(28), 12538–12545.

[39] Guo, X. M., Guo, H. D., Fu, L. S., Zhang, H., Carlos, L. D., Deng, R. P., Yu, J. B. (2008). Synthesis and photophysical properties of novel organic–inorganic hybrid materials covalently linked to a europium complex. J. Photochem. Photobiol., A, 200(2–3), 318–324.

[40] Guo, X. M., Wang, X. M., Zhang, H. J., Fu, L. S., Guo, H. D., Yu, J. B., Carlos, L. D., Yang, K. Y. (2008). Preparation and luminescence properties of covalent linking of luminescent ternary europium complexes on periodic mesoporous organosilica. Micropor. Mesopor. Mater., 116 (1–3), 28–35.

[41] Hao, J. N., Yan, B. (2014). Highly sensitive and selective fluorescent probe for Ag^+ based on a Eu^{3+} post-functionalized metal–organic framework in aqueous media. J. Mater. Chem. A, 2, 18018–18025.

[42] Hao, J. N., Yan, B. (2015). A water-stable lanthanide-functionalized MOF as a highly selective and sensitive fluorescent probe for Cd^{2+}. Chem. Comm., 51, 7737–7740.

[43] Harrison, B. S., Foley, T. J., Knefely, A. S., Mwaura, J. K., Cunningham, G. B., Kang, T. S., Bouguettaya, M., Boncella, J. M., Reynolds, J. R., Schanze, K. S. (2004). Near-Infrared photo- and electroluminescence of alkoxy-substituted poly(p-phenylene) and nonconjugated polymer/lanthanide tetraphenylporphyrin blends. Chem. Mater., 16(15), 2938–2947.

[44] Huang, X., Wang, Q., Yan, X., Xu, J., Liu, W., Wang, Q., Tang, Y. (2011). Encapsulating a ternary europium complex in a silica/polymer hybrid matrix for high performance luminescence application. J. Phys. Chem. C, 115(5), 2332–2340.

[45] Huang, Y., Lei, J., Cheng, Y., Ju, H. (2016). Ratiometric electrochemiluminescent strategy regulated by electrocatalysis of palladium nanocluster for immunosensing. Biosens. Bioelectron., 77, 733–739.

[46] Judd, B. R. (1962). Optical absorption intensities of rare-earth. Ions Phys. Rev, 127, 750.

[47] Justel, T., Nikol, H., Ronda, C. (1998). New developments in the field of luminescent materials for lighting and display. Angew. Chem. Int. Ed., 37(22), 3084–3103.

[48] Kido, J., Okamoto, Y. (2002). Organo lanthanide metal complexes for electroluminescent material. Chem. Rev., 102(6), 2357–2368.

[49] Klink, S. I., Hebbink, G. A., Grave, L., Van Veggel, F. C. J. M., Reinhoudt, D. N., Slooff, L. H., Polman, A., Hofstraat, J. W. (1999). Sensitized near-infrared luminescence from polydentate triphenylene-functionalized Nd^{3+}, Yb^{3+}, and Er^{3+} complexes. J. Appl. Phys., 86(3), 1181.

[50] Kuriki, K., Koike, Y., Okamoto, Y. (2002). Plastic optical fiber lasers and amplifiers containing lanthanide complexes. Chem. Rev., 102(6), 2347–235.

[51] Kwon, B. H., Jang, H. S., Yoo, H. S., Kim, S. W., Kang, D. S., Maeng, S., Jang, D. S., Kimand, H., Jeon, D. Y. (2011). White-light emitting surface-functionalized ZnSe quantum dots: europium complex-capped hybrid nanocrystal. J. Mater. Chem., 21, 12812–12818.

[52] Li, H., Liu, R., P.,, Wang, Y. G., Zhang, L., Yu, J. B., Zhang, H. J., Liu, B. Y., Schubert, U. (2009). Preparation and luminescence properties of hybrid titania immobilized with lanthanide complexes. J. Phys. Chem. C, 113(10), 3945–3949.

[53] Li, H. R., Lin, J., Zhang, H. J., Fu, L. S., Meng, Q. G., Wang, S. B. (2002). Preparation and luminescence properties of hybrid materials containing europium(III) complexes covalently bonded to a silica matrix. Chem. Mater., 14(9), 3651–3655.

[54] Li, H. R., Lin, J., Zhang, H. J., Li, H. C., Fu, L. S., Meng, Q. G. (2001). Novel, covalently bonded hybrid materials of europium (terbium) complexes with silica. Chem Commun., 1212–1213.

[55] Li, H. R., Yu, J. B., Liu, F. Y., Zhang, H. J., Fu, L. S., Meng, Q. G., Peng, C. Y., Lin, J. (2004). Preparation and luminescence properties of in situ formed lanthanide complexes covalently grafted to a silica network. New J. Chem., 28, 1137–1141.

[56] Li, P., Zhang, Y. Z., Wang, Y. G., Wang, Y. J., Li, H. R. (2014). Luminescent europium(iii)–β-diketonate complexes hosted in nanozeolite L as turn-on sensors for detecting basic molecules. Chem. Comm., 50, 13680–13682.

[57] Li, Y., Yan, B. (2010). Lanthanide (Tb^{3+}, Eu^{3+}) functionalized MCM-41 through modified meta-aminobenzoic acid linkage: Covalently bonding assembly, physical characterization and photoluminescence. Micropor. Mesopor. Mater., 128(1–3), 62–70.

[58] Li, Y., Yan, B., Li, Y. J. (2010). Sulfide functionalized lanthanide (Eu/Tb) periodic mesoporous organosilicas (PMOs) hybrids with covalent bond: Physical characterization and photoluminescence. Micropor. Mesopor. Mater., 132(1–2), 87–93.

[59] Li, Y., Yan, B., Yang, H., Construction, C. (2008). Photoluminescence of mesoporous hybrids containing europium(III) complexes covalently bonded to SBA-15 directly functionalized by modified β-Diketone. J. Phys. Chem. C, 112(3), 3959–3968.

[60] Li, Y. J., Wang, L., Yan, B. (2011). Photoactive lanthanide hybrids covalently bonded to functionalized periodic mesoporous organosilica (PMO) by calix[4]arene derivative. J. Mater. Chem., 21, 1130–1138.

[61] Li, Y. J., Yan, B. (2009). Lanthanide (Eu^{3+}, Tb^{3+})/β-Diketone Modified Mesoporous SBA-15/organic polymer hybrids: chemically bonded construction, physical characterization, and photophysical properties. Inorg. Chem., 48(17), 8276–8285.

[62] Li, Y. J., Yan, B. (2010). Photophysical Properties of a Novel Organic–Inorganic Hybrid Material: Eu(III)-β-Diketone Complex Covalently Bonded to SiO_2/ZnO Composite Matrix. Photochem. Photobiol., 86, 1008–1015.

[63] Li, Y. J., Yan, B. (2010). Hybrid materials of SBA-16 functionalized by rare earth (Eu^{3+}, Tb^{3+}) complexes of modified β-diketone (TTA and DBM): Covalently bonding assembly and photophysical properties. J. Solid State Chem., 183(4), 871–877.

[64] Li, Y. J., Yan, B. (2010). Photoactive europium centered mesoporous hybrids with 2-thenoyltrifluoroacetone functionalized SBA-16 and organic polymers. Dalton Trans., 39, 2554–2562.

[65] Li, Y. J., Yan, B., Li, Y. (2010). Lanthanide (Eu^{3+}, Tb^{3+}) Centered Mesoporous Hybrids with 1,3-Diphenyl-1,3-Propanepione Covalently Linking SBA-15 (SBA-16) and Poly(methylacrylic acid). Chem.–Asian J., 5(7), 1642–1651.

[66] Li, Y. J., Yan, B., Wang, L. (2011). Rare earth (Eu^{3+}, Tb^{3+}) mesoporous hybrids with calix[4] arene derivative covalently linking MCM-41: Physical characterization and photoluminescence property. J. Solid State Chem., 184(9), 2571–2575.

[67] Li, Y. Y., Yan, B., Guo, L. (2011). Photofunctional ternary rare earth (Eu^{3+}, Tb^{3+}, and Sm^{3+}) hybrid xerogels with hexafluoroacetylacetonate derived building block and bis(2-methoxyethyl)ether through coordination bonds. Inorg. Chem. Commun., 14(6), 910–912.

[68] Lian, X., Yan, B. (2016). Novel core–shell structure microspheres based on lanthanide complexes for white-light emission and fluorescence sensing. Dalton Trans., 45, 2666–2673.

[69] Lima, P. P., Ferreira, R. A. S., Ju´nior, S. A., Malta, O. L., Carlos, L. D. (2009). Terbium(III)-containing organic–inorganic hybrids synthesized through hydrochloric acid catalysis. J. Photochem. Photobiol., A, 201(2–3), 214–221.

[70] Liu, C., Yan, B. (2015). Highly effective chemosensor of a luminescent silica@lanthanide complex@MOF heterostructured composite for metal ion sensing. RSC Adv, 5, 101982–101988.

[71] Liu, C., Yan, B. (2016). A novel photofunctional hybrid material of pyrene functionalized metal-organic framework with conformation change for fluorescence sensing of Cu^{2+}. Sens. Actuators B Chem., 235(1), 541.

[72] Liu, F. Y., Carlos, L. D., Ferreira, R. A. S., Rocha, J., Ferro, M. C., Tourrette, A., Quignard, F., Robitzer, M., Synthesis, T. (2010). Photoluminescence of Lanthanide-Containing Chitosan–Silica Hybrids. J. Phys. Chem. B, 114(1), 77–83.

[73] Liu, F. Y., Fu, L. S., Wang, J., Meng, Q. G., Li, H. R., Guo, J. F., Zhang, H. J. (2003). Luminescent film with terbium-complex-bridged polysilsesquioxanes. New J. Chem., 27, 233–235.

[74] Liu, J. L., Yan, B. (2008). Lanthanide (Eu^{3+}, Tb^{3+}) Centered Hybrid Materials using Modified Functional Bridge Chemical Bonded with Silica: Molecular Design, Physical Characterization, and Photophysical Properties. J. Phys. Chem. B, 112(35), 10898–10907.

[75] Liu, J. L., Yan, B. (2010). Rare-earth (Eu^{3+}, Tb^{3+}) hybrids through amide bridge: Chemically bonded self-assembly and photophysical properties. J. Organomet. Chem., 695(4), 580–587.

[76] Liu, J. L., Yan, B. (2011). Lanthanide-centered organic–inorganic hybrids through a functionalized aza-crown ether bridge: coordination bonding assembly, microstructure and multicolor luminescence. Dalton Trans., 40, 1961–1968.

[77] Lu, H. F., Yan, B., Liu, J. L. (2009). Functionalization of Calix[4]arene as a Molecular Bridge To Assemble Luminescent Chemically Bonded Rare-Earth Hybrid Systems. Inorg. Chem., 48(9), 3966–3975.

[78] Lunstroot, K., Driesen, K., Nockemann, P., Van Hecke, K., Van Meervelt, L., Goerller-Walrand, C., Binnemans, K., Bellayer, S., Viau, L., Le Bideau, J., Vioux, A. (2009). Lanthanide-doped luminescent ionogels. Dalton Trans., 298–306.

[79] Luttmer, J. D., Bard, A. J. (1981). Electrogenerated chemiluminescence. 38. Emission intensity-time transients in the tris(2,2'-bipyridine)ruthenium(II) system. J. Phys. Chem., 85(9), 1155–1159.

[80] Ma, H., Zhou, J., Li, Y., Han, T., Zhang, Y., Hu, L., Du, B., Wei, Q. (2016). A label-free electrochemiluminescence immunosensor based on $EuPO_4$ nanowire for the ultrasensitive detection of Prostate specific antigen. Biosens. Bioelectron., 80, 352–358.

[81] Maggini, L., Mohanraj, J., Traboulsi, H., Parisini, A., Accorsi, G., Armaroli, N., Bonifazi, D. (2011). A Luminescent Host–Guest Hybrid between a Eu^{III} Complex and MWCNTs. Chem.–Eur. J., 17(31), 8533–8537.

[82] Melhuish, W. H. (1984). VI: Molecular luminescence spectroscopy. Pure Appl. Chem., 56(2), 231–245.
[83] Mesquita, M. E., Nobre, S. S., Fernandes, M., Ferreira, R. A. S., Santos, S. C. G., Rodrigues, M. O., Carlos, L. D., Bermudezc, V. D. (2009). Highly luminescent di-ureasil hybrid doped with a Eu (III) complex including dipicolinate ligands. J. Photochem. Photobiol., A, 205(2–3), 156–160.
[84] Minoofar, P. N., Hernandez, R., Chia, S., Dunn, B., Zink, J. I., Franville, A. C. (2002). Placement and characterization of pairs of luminescent molecules in spatially separated regions of nanostructured thin films. J. Am. Chem. Soc., 124(48), 14388–14396.
[85] Molina, C., Ferreira, R. A. S., Poirier, G., Fu, L., Ribeiro, S. J. L., Messsaddeq, Y., Carlos, L. D. (2008). Er^{3+}-based diureasil organic–inorganic hybrids. J. Phys. Chem. C, 112(49), 19346–19352.
[86] Ofelt, G. S. (1962). Intensities of crystal spectra of rare-earth. Ions J. Chem. Phys., 37, 511.
[87] Okmoto, Y., Ueba, Y., Zhanibekov, N. F. D., Bank, E. (1981). Rare earth metal containing polymers. Characterisation of ion containing polymers structure using rare earth metal fluorescence., Macromolecules, 14, 17.
[88] Ozawa, L., Itoh, M. (2003). Cathode ray tube phosphors. Chem. Rev., 103(10), 3832–3856.
[89] Ozin, G. A., Arsenault, A. C. (2006). Nanochemistry – A Chemical Approach to Nanomaterials. 2nd edition, Royal Society of Chemistry, Cambridge, U.K.
[90] Parra, D. F., Brito, H. F., Matos, J. D. R., Carlos, L. D. (2002). Enhancement of the luminescent intensity of the novel system containing Eu^{3+}-β-diketonate complex doped in the epoxy resin. J. Appl. Polym. Sci., 83(12), 2716–2726.
[91] Polman, A., Veggel, F. C. J. M. (2004). Van, Broadband sensitizers for erbium-doped planar optical amplifiers: review. J. Opt. Soc. Am. B, 21(5), 871–892.
[92] Qiao, X. F., Yan, B. (2008). Assembly, characterization, and photoluminescence of hybrids containing europium(III) complexes covalently bonded to inorganic Si–O Networks/Organic polymers by modified β-diketone. J. Phys. Chem. B, 112(47), 14742–14750.
[93] Qiao, X. F., Yan, B. (2008). Molecular construction and photophysics of luminescent covalently bonded hybrids by grafting the lanthanide ions into the silicon–oxygen networks and carbon chain. J. Photochem. Photobiol., A, 199(2–3), 188–196.
[94] Qiao, X. F., Yan, B. (2009). Chemically bonded assembly and photophysical properties of luminescent hybrid polymeric materials embedded into silicon–oxygen network and carbon unit. J. Organomet. Chem., 694(20), 3232–3241.
[95] Qiao, X. F., Yan, B. (2009). Covalently bonded assembly of lanthanide/silicon–oxygen network/polyethylene glycol hybrid materials through functionalized 2-thenoyltrifluoroacetone linkage. J. Phys. Chem. B, 113(35), 11865–11875.
[96] Qiao, X. F., Yan, B. (2009). Hybrid materials of lanthanide centers/functionalized thenoyltrifluoroacetone/silicon–oxygen network/polymeric chain: coordination bonded assembly, physical characterization, and photoluminescence. Inorg. Chem., 48(11), 4714–4723.
[97] Qiao, X. F., Zhang, H. Y., Yan, B. (2010). Photoactive binary and ternary lanthanide (Eu^{3+}, Tb^{3+}, Nd^{3+}) hybrids with p-tert-butylcalix[4]arene derived Si–O linkages and polymers. Dalton Trans., 39, 8882–8892.
[98] Richter, M. M. (2004). Electrochemiluminescence (ECL). Chem. Rev., 104(6), 3003–3036.
[99] Richter, M. M., Bard, A. J. (1996). Electrogenerated chemiluminescence. 58. Ligand-sensitized electrogenerated chemiluminescence in europium labels. Anal. Chem., 68(15), 2641.
[100] Sabbatini, N., Guardigli, M., Lehn, J. M. (1993). Luminescent lanthanide complexes as photochemical supramolecular devices. Coord. Chem. Rev., 123(1–2), 201–228.
[101] Sabbatini, N., Mecati, A., Guardigli, M., Balazani, V., Lehn, J. M., Zeissel, R., Ungaro, R. (1991). Lanthanide luminescence in supramolecular species. J. Lumin., 48–49(2), 463–468.
[102] Shen, J., Zhou, T., Huang, R. (2019). Recent advances in electrochemiluminescence sensors for pathogenic bacteria detection. Micromachines, 10(8), 532.

[103] Shen, X., Yan, B. (2015). Photofunctional hybrids of lanthanide functionalized bio-MOF-1 for fluorescence tuning and sensing. J. Coll. Interf. Sci., 451, 63–68.
[104] Shen, X., Yan, B. (2016). Barcoded materials based on photoluminescent hybrid system of lanthanide ions-doped metal organic framework and silica via ion exchange. J. Coll. Interf. Sci., 468, 220–226.
[105] Snitzer, E., Young, G. (1968). Glass Lasers. Marcel Dekker, New York.
[106] Sui, Y. L., Yan, B. (2006). Fabrication and photoluminescence of molecular hybrid films based on the complexes of 8-hydroxyquinoline with different metal ions via sol–gel process. J. Photochem. Photobiol., A, 182(1), 1–6.
[107] Sun, L. N., Dang, S., Yu, J. B., Feng, J., Shi, L. Y., Zhang, H. J. (2010). Near-Infrared luminescence from visible-light-sensitized hybrid materials covalently linked with Tris(8-hydroxyquinolinate)-lanthanide [Er(III), Nd(III), and Yb(III)] derivatives. J. Phys. Chem. B, 114 (49), 16393–16397.
[108] Sun, L. N., Zhang, H. J., Fu, L. S., Liu, F. Y., Meng, Q. G., Peng, C. Y., Liu, J. B., New Sol–Gel, A. (2005). Material doped with an erbium complex and its potential optical-amplification application. Adv. Funct. Mater., 15(5), 1041–1048.
[109] Sun, N. N., Yan, B. (2016). Lanthanide complex inside–outside double functionalized zeolite A hybrid materials for luminescence sensing. New J. Chem., 40, 6924–6930.
[110] Sun, N. N., Yan, B. (2017). Rapid and facile ratiometric detection of CO_3^{2-} based on heterobimetallic metal-organic frameworks (Eu/Pt-MOFs). Dyes Pigm., 42, 1–7.
[111] Tan, C. L., Wang, Q. M. (2011). Reversible terbium luminescent polyelectrolyte hydrogels for detection of $H_2PO_4^-$ and HSO_4^- in water. Inorg. Chem., 50(7), 2953–2956.
[112] Taniguchi, H., Kido, J., Nishiya, M., Sasaki, S. (1995). Europium chelate solid laser based on morphology-dependent resonances. Appl. Phys. Lett., 67, 1060.
[113] Ueba, Y., Banks, E., Okmoto, Y. (1980). Investigation on the synthesis and characterization of rare earth metal-containing polymers. II. Fluorescence properties of Eu^{3+}–polymer complexes containing β-diketone ligand. J. Appl. Polym. Sci., 25(9), 2007–2017.
[114] Valenti, G., Fiorani, A., Li, H., Sojic, N., Paolucci, F. (2016). Essential role of electrode materials in electrochemiluminescence applications. ChemElectroChem., 3(12), 1990–1997.
[115] Van Deun, R., Fias, P., Binnemans, K., Gorller-Walrand, C. (2003). Halogen substitution as an efficient tool to increase the near-infrared photoluminescence intensity of erbium(iii) quinolinates in non-deuterated DMSO. Phys. Chem. Chem. Phys., 5, 2754–2757.
[116] Wang, C., Yan, B. (2011). Photophysical properties of rare earth (Eu^{3+}, Sm^{3+}, Tb^{3+}) complex covalently immobilized in hybrids Si-O-B xerogels. J. Fluoresc., 21(3), 1239–1247.
[117] Wang, C., Yan, B., Liu, J. L., Guo, L. (2011). Photoactive europium hybrids of β-diketone-modified polysilsesquioxane bridge linking Si–O–B(Ti)–O xerogels. Eur. J. Inorg. Chem., 6, 879–887.
[118] Wang, J. X., Zhuo, Y., Zhou, Y., Wang, H. J., Yuan, R., Chai, Y. Q. (2016). Ceria doped zinc oxide nanoflowers enhanced luminol-based electrochemiluminescence immunosensor for amyloid-β detection. ACS Appl. Mater. Interfaces, 8(20), 12968–12975.
[119] Wang, Q. M., Tan, C. L., Cai, W. S. (2012). A targetable fluorescent sensor for hypochlorite based on a luminescent europium complex loaded carbon nanotube. Analyst, 137, 1872–1875.
[120] Wang, Q. M., Yan, B. (2004). Novel luminescent terbium molecular-based hybrids with modified meta-aminobenzoic acid covalently bonded with silica. J. Mater. Chem., 14, 2450–2454.
[121] Wang, Q. M., Yan, B. (2006). Molecular assembly of red and green nanophosphors from amine-functionalized covalent linking hybrids with emitting centers of Eu^{3+} and Tb^{3+} ions. J. Photochem. Photobiol., A, 178(1), 70–75.

[122] Wang, Q. M., Yan, B. (2006). Terbium/zinc luminescent hybrid siloxane-oxide materials bridged by novel ureasils linkages. J. Organomet. Chem., 691(4), 545–550.
[123] Wang, Q. M., Yan, B. (2007). Molecular assembly of novel luminescent terbium hybrid materials with modified 5-ethylpyridine-2,3-dicarboxylic acid as a functional bridge through an in situ sol–gel process. Opt. Mater., 29(5), 510–515.
[124] Wang, Q. M., Yan, B. (2008). Covalently bonded assembly and photoluminescent properties of rare earth/silica/poly(methyl methacrylate-co-maleic anhydride) hybrid materials. J. Photochem. Photobiol., A, 197(2-3), 213–219.
[125] Wang, T., Liu, M., JiQ., Wang, Y. (2015). RSC Adv, 5, 103433–103438.
[126] Wang, T. R., Li, P., Li, H. R. (2014). Color-tunable luminescence of organoclay-based hybrid materials showing potential applications in white LED and thermosensors. ACS Appl. Mater. Interf., 6(15), 12915–12921.
[127] Wang, X. L., Yan, B., Liu, J. L. (2010). Photoactive rare earth (Eu^{3+}, Tb^{3+}) hybrid with organically modified silica covalently bonded polymeric chain. Colloid Polym. Sci., 288, 1139–1150.
[128] Wang, Y., Li, H. R., Feng, Y., Zhang, H. J., Calzaferri, G., Ren, T. Z. (2010). Orienting zeolite l microcrystals with a functional linker. Angew. Chem., Int. Ed., 49(8), 1434–1438.
[129] Wang., Y. G., Wang, L., Li, H. R., Liu, P., Qin, D. S., Liu, B. Y., Zhang, W. J., Deng, R. P., Zhang, H. J. (2008). Synthesis and luminescence properties of hybrid organic–inorganic transparent titania thin film activated by in-situ formed lanthanide complexes. J. Solid State Chem., 181 (3), 562–566.
[130] Weissman, S. I. (1942). Intramolecular energy transfer the fluorescence of complexes of europium. J. Chem. Phys., 10, 214.
[131] Weng, H., Yan, B. (2016). Lanthanide coordination polymers for multi-color luminescence and sensing of Fe^{3+}. Inorg. Chem. Comm., 63, 11–15.
[132] Weng, H., Yan, B. (2017). A Eu(III) doped metal-organic framework conjugated with fluorescein-labeled single stranded DNA for detection of Cu(II) and sulfide. Anal. Chim. Acta, 988, 89–95.
[133] Werts, M. H. V. (2005). Making sense of lanthanide luminescence. Sci. Pro., 88(2), 101–131.
[134] White, K. A., Chengelis, D. A., Gogick, K. A., Stehman, J., Rosi, N. L., Petoud, S. (2009). Near-infrared luminescent lanthanide MOF barcodes. J. Am. Chem. Soc., 131(50), 18069–18071.
[135] Wolfbeis, O. S., Durkop, A., Wu, M., Lin, Z., Europium-ion-based luminescent, A. (2002). Sensing probe for hydrogen peroxide. Angew. Chem. Int. Ed., 41(23), 4495–4498.
[136] Wu, D., Xin, X., Pang, X., Pietraszkiewicz, M., Hozyst, R., Sun, X., Wei, Q. (2015). Application of europium multiwalled carbon nanotubes as novel luminophores in an electrochemiluminescent aptasensor for thrombin using multiple amplification strategies. ACS Appl. Mater. Interfaces, 7(23), 12663–12670.
[137] Wybourne, B. G. (1965). Spectroscopic Properties of the Rare Earths. New York, Wiley.
[138] Xu, X., Qin, X., Wang, L., Wang, X., Lu, J., Qiu, X., Zhu, Y. (2019). Lanthanide terbium complex: synthesis, electrochemiluminescence (ECL) performance, and sensing application. Analyst, 144, 2359–2366.
[139] Xu, X. Y., Yan, B. (2015). Eu(III)-functionalized MIL-124 as fluorescent probe for highly selectively sensing ions and organic small molecules especially for Fe(III) and Fe(II). ACS Appl. Mater. Interf., 7(1), 721–729.
[140] Xu, X. Y., Yan, B. (2016). Eu(III) functionalized Zr-based metal-organic framework as excellent fluorescent probe for Cd^{2+} detection in aqueous environment. Sens. Actuators B Chem., 222, 347–353.
[141] Xu, X. Y., Yan, B. (2016). Fabrication and application of a ratiometric and colorimetric fluorescent probe for Hg^{2+} based on dual-emissive metal–organic framework hybrids with carbon dots and Eu^{3+}. J. Mater. Chem. C, 4, 1543–1549.

[142] Yan, B. (2017). Photofunctional rare earth hybrid materials. Springer Ser. Mater. Sci., 251, 199.
[143] Yan, B., Kong, L. L., Zhou, B. (2009). A luminescent covalently bonded rare earth hybrid material by functionalized trifluoroacetylacetone linkage. J. Non-Cryst. Solids, 355(22–23), 1281–1284.
[144] Yan, B., Li, Y. (2010). Luminescent ternary inorganic-organic mesoporous hybrids Eu(TTASi-SBA-15) phen: covalent linkage in TTA directly functionalized SBA-15. Dalton Trans., 39, 1480–1487.
[145] Yan, B., Lu, H. F. (2009). Lanthanide-centered inorganic/organic hybrids from functionalized 2-pyrrolidinone-5-carboxylic acid bridge: Covalently bonded assembly and luminescence. J. Organomet. Chem., 694(16), 2597–2603.
[146] Yan, B., Lu, H. F. (2009). Novel leaf-shaped hybrid micro-particles: chemically bonded self-assembly, microstructure and photoluminescence. J. Photochem. Photobiol. A, 205(2–3), 122–128.
[147] Yan, B., Ma, D. J. (2006). From chemistry to materials, design and photophysics of functional terbium molecular hybrids from assembling covalent chromophore to alkoxysilanes through hydrogen transfer addition. J. Solid State Chem., 179(7), 2059–2066.
[148] Yan, B., Qiao, X. F. (2007). Photophysical properties of terbium molecular-based hybrids assembled with novel ureasil linkages. Photochem. Photobiol., 83(4), 971–978.
[149] Yan, B., Qiao, X. F. (2007). Rare-earth/inorganic/organic polymeric hybrid materials: molecular assembly, regular microstructure and photoluminescence. J. Phys. Chem. B, 111 (43), 12362–12374.
[150] Yan, B., Sui, Y. L., Liu, J. L. (2009). Photoluminescent hybrid thin films fabricated with lanthanide ions covalently bonded silica. J. Alloys Compd., 476(1–2), 826–829.
[151] Yan, B., Wang, C. (2011). Rare earth (Eu^{3+}, Tb^{3+}) centered composite gels Si-O-M (M = B, Ti) through hexafluoroacetyl-acetone building block: Sol–gel preparation, characterization and photoluminescence. Mater. Res. Bull., 46(12), 2515–2522.
[152] Yan, B., Wang, C., Guo, L., Liu, J. L. (2010). Photophysical Properties of Eu(III) Center Covalently Immobilized in Si-O-B and Si-O-Ti Composite Gels. Photochem. Photobiol., 86(3), 499–507.
[153] Yan, B., Wang, F. F. (2007). Molecular design and photo-physics of quaternary hybrid terbium centered systems with novel functional di-urea linkages of strong chemical bonds through hydrogen transfer addition. J. Organomet. Chem., 692(12), 2395–2401.
[154] Yan, B., Wang, Q. M. (2008). Two luminescent molecular hybrids composed of bridged Eu(iii)-β-diketone chelates covalently trapped in silica and titanate gels. Cryst. Growth Des., 8(5), 1484–1489.
[155] Yan, B., Zhao, L. M., Liu, J. L. (2008). Molecular assembly and photophysical properties of covalently bonded rare earth polymeric hybrid materials phen–RE–MSMA (MS). J. Photochem. Photobiol., A, 199(1), 50–56.
[156] Yan, B., Zhao, L. M., Wang, X. L., Zhao, Y. (2011). Sol-gel preparation, microstructure and luminescence of rare earth/silica/polyacrylamide hybrids through double functionalized covalent Si–O linkage. RSC Adv., 1, 1064–1071.
[157] Yan, B., Zhao, Y., Li, Q. P. (2011). europium hybrids/SiO_2/semiconductor: multi-component sol–gel composition, characterization and photoluminescence. J. Photochem. Photobiol., A, 222(2–3), 351–359.
[158] Yan, B., Zhou, B. (2008). Two photoactive lanthanide (Eu^{3+}, Tb^{3+}) hybrid materials of modified β-diketone bridge directly covalently bonded mesoporous host (MCM-41). J. Photochem. Photobiol., A, 195(2–3), 314–322.
[159] Yan, B., Zhou, B., Li, Y. (2009). Covalently bonding assembly and photophysical properties of luminescent molecular hybrids Eu–TTA–Si and Eu–TTASi–MCM-41 by modified thenoyltrifluoroacetone. Micropor. Mesopor. Mater., 120(3), 317–324.

[160] Yan Wang, Q. M. B. (2006). Molecular assembly of red and green nanophosphors from amine-functionalized covalent linking hybrids with emitting centers of Eu^{3+} and Tb^{3+} ions. J. Photochem. Photobiol. A, 178(1), 70–75.

[161] Yang, Y., Zhang, Y., Shu, G., Dong, Q., Zou, L., Zhu, Y. (2015). Electrochemiluminescence properties of Tb(III) nicotinic acid complex and its analytical application. J. Lumin., 159, 73–78.

[162] Yao, Y. L., Wang, Y. G., Li, Z. Q., Li, H. R. (2015). Reversible on–off luminescence switching in self-healable hydrogels. Langmuir, 31(46), 12736–12741.

[163] Zhao, L. M., Yan, B. (2005). A novel path to luminescent hybrid molecular materials: modifying the hydroxyl group of 6-hydroxynicotinic acid by grafting to a silica network. Appl. Organomet. Chem., 19(9), 1060–1064.

[164] Zhou, Y., Chen, H. H., Yan, B. (2014). An Eu^{3+} post-functionalized nanosized metal–organic framework for cation exchange-based Fe^{3+}-sensing in an aqueous environment. J. Mater. Chem. A, 2, 13691–13697.

[165] Zhou, Y., Yan, B. (2015). Ratiometric multiplexed barcodes based on luminescent metal–organic framework films. J. Mater. Chem. C, 3, 8413–8418.

[166] Zhou, Y., Yan, B. (2016). A responsive MOF nanocomposite for decoding volatile organic compounds. Chem. Comm., 52, 2265–2268.

[167] Zhou, Z., Wang, Q. M., Zeng, Z., Yang, L. T., Ding, X. P., Lin, N., Cheng, Z. S. (2013). Polyurethane-based Eu(III) luminescent foam as a sensor for recognizing Cu^{2+} in water. Anal. Methods, 5, 6045–6050.

[168] Zhou, Z., Zheng, Y. H., Wang, Q. M. (2014). Extension of novel lanthanide luminescent mesoporous nanostructures to detect fluoride. Inorg Chem, 53(3), 1530–1536.

Sivakumar Vaidyanathan, Priyadarshini Pradhan
Chapter 2
Lanthanide-Based Molecular Thermometers: An Overview

Abstract: Metal organic frameworks (MOFs) are amazing platforms for building luminescent properties as their elementary units, metal ions, organic linkers and guest ions or molecules are on the whole possible wellsprings of light emission. Temperature is one of the most significant physical properties influencing dynamics and suitability of natural and designed frameworks. Since the luminescence of certain rare-earth-based MOFs changes impressively with temperature, in the last scarcely any years, these materials have been investigated as optical thermometers, particularly in temperature sensing based on the intensity proportions of two distinct electronic transitions. In this chapter, we mainly emphasize the important concepts and ideas that help the structure of such ratiometric thermometers. The detailed investigations on their crystal structure, synthesis and temperature-dependent photoluminescence properties are briefly discussed. The improved optical properties because of the energy transference process between the organic linkers and trivalent rare earth ions (Eu^{3+}, Tb^{3+}) as well as metal–metal ions make Ln-MOFs as potential candidates to act as excellent temperature sensors. In this book, we have tried our best to correlate structure–composition–property of Ln-MOFs with special emphasis on temperature sensors.

Keywords: Lanthanide metal organic framework, ratiometric, luminescent thermometer, dual emission, sensitivity

2.1 Introduction

Temperature is a significant and very influential physical parameter in the standard of living, scientific research and industrial manufacturing [46, 50]. Simultaneously, temperature likewise assumes a critical role in the procedures of physiology and pathology research. For instance, temperature is not just a crucial factor affecting the cell division and determining the growth rate of tissue, yet additionally affecting numerous biological processes inside the cells, including protein synthesis, ion transport, gene expression, neural signaling, and cellular respiration [50]. Consequently, precise estimation of temperature is of extraordinary importance. Essential advancement in thermometers depends on contact thermometers, counting thermocouples, mercury thermometers, thermistors and bimetallic thermometers, which predominantly measure temperature by the variations of volume, electric potential and conductance. The contact temperature sensors need heat transmission throughout usage and involve adequate time to accomplish the condition of warm balance between the temperature sensors and the test specimen. The nature of interaction gives rise to contact temperature sensors that are not reasonable for some extraordinary conditions; for example, in liquids, cells/vivo and the surface of quick moving item [50]. Also, such customary sensors have a few shortcomings; for example, their obtrusiveness, huge size and utilization of electrical modules. So they are not relevant for quick developing of potential submicrometer objects or strong electromagnetic fields. To defeat previously mentioned confinements, numerous works have been done using noncontact temperature detectors, which includes luminous temperature sensors, and considerable advancement has been made in the field of noncontact thermometers [7, 58]. In any case, the fly in the ointment is that the temperature estimations of starting luminous temperature detectors depend on single enlightening center and are liable to outside circumstances counting material concentration and homogeneousness, excitation power, atmosphere prompted nonradiative relaxation and the drifts of the optoelectronic framework (e.g., lights and sensors) [50]. Thus, ratiometric luminescence was structured, so as to deal with the above nearness issues, which utilizes the intensity proportion of two illuminating/luminescence centers (e.g., Er^{3+}/Yb^{3+}, Tb^{3+}/Eu^{3+}, Yb^{3+}/Nd^{3+}) as the measuring parameter [50]. It is important to note that these types of thermometers need not bother the contact with the entity, are self-regulating, are profoundly precise, have a quick response and an outstanding resolution (at the submicrometer level), and are appropriate under most electromagnetic circumstances. Despite the fact that ratiometric temperature sensors can be based on organic molecules, polymer dots, chelate complexes, up-conversion nanoparticles and others [31, 33, 42, 58], trivalent lanthanide (Ln^{3+})-based metal organic frameworks (Ln-MOFs) have stood out as a result of their interesting luminescent properties [14, 38, 58]. Ln-MOFs, self-amassed from lanthanide comprising clusters with organic ligands, are flexible and have steady luminous procedures with high outflow quantum yields (QY) [1, 9, 13, 39, 58]. They consolidate the spectroscopic characteristics of rare earths (e.g., extensive emission range, distinct fine band and long fluorescence lifetime) with the exclusive superiority of metal organic frameworks (MOFs), for example, assorted frameworks, everlasting porosity and extraordinary designability [26, 28, 30, 41, 58, 61]. They have hence pulled in an incredible arrangement of considerations in the previous scarcely any years and opened another course for creating novel luminescent sensory materials [52, 58]. They are flexible multifunctional luminous components and have been examined for an extensive scope of detecting applications [9, 18, 27, 30, 55, 59], particularly in temperature estimation. The luminescent properties of various rare earth ions in the multicenter Ln-MOFs are firmly dependent on temperature. Therefore, the proportion of luminescence intensities of two distinct centers can be utilized to figure it out by self-adjusting luminous temperature sensors of high exactness and quick response [59].

Recently, numerous researchers dedicated their research direction in the noncontact temperature sensor. In this chapter, efforts have been made to review the lanthanide-based ratiometric temperature sensing, including the fundamental aspects.

Basically, all MOF temperature detectors discover the outright temperature through the estimation of intensities of two electronic transitions of particular radiating centers (genuine double center thermometers): an organic linker and a lanthanide ion (Eu^{3+} or Tb^{3+}), two Ln^{3+} ions (prior to Eu^{3+} and Tb^{3+}) or a dye fascinated in the MOF nanopores, and an Ln^{3+} ion (Figure 2.1) [16]. The bimetallic Ln-MOF temperature sensors for numerous temperature regions, for example, cryogenic ($T < 100$ K) [16, 29, 37, 59], medium ($T = 100–300$ K) [21, 56], physiological ($T = 293–323$ K) [8, 63] and high ($T > 373$ K) [49, 57] were presented.

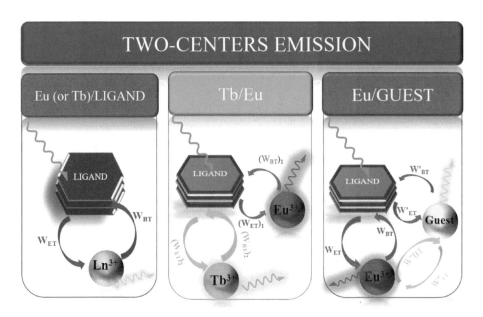

Figure 2.1: Schematic portrayal of the three types of dual-center emission MOF temperature sensors. Center and left panels: Thermally activated energy transference between host and metal ion administrate the changes in the thermometric parameters, where W_{ET} is the rate of energy transference from ligand to metal ion and W_{BT} is the rate of energy back transference from metal to ligand [$W_{BT} = W_{ET}\exp(-\Delta E/K_B T)$]. Right panel: Mechanisms of energy transference functioning in the only thermometer reported so far comprising a perylene molecule hosted in the MOFs nanopores [10, 38]. Energy transference between ion–ion is not depicted. Redrawn from [38].

2.2 Energy Transfer Mechanism

The schematic representation of energy transfer between ligand (organic antenna) and lanthanide ions is shown in Figure 2.2. The precise control of energy-level location is very much crucial for effectual energy transference from ligand to Ln^{3+} ion.

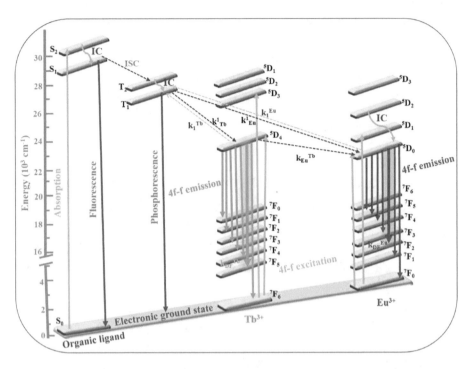

Figure 2.2: Jablonski diagram describing the energy transference and back transference from ligand-to-metal and metal-to-metal ions.

The luminescence of the rare-earth-based MOFs, which is dependent on temperature, is controlled by the energy transference method, between ligand-to-Tb^{3+} and ligand-to-Eu^{3+} along with Tb^{3+}-to-Eu^{3+} transference, respectively. The emissions of both the trivalent lanthanide ions, that is, Tb^{3+} and Eu^{3+} can be sensitized by the organic linker (which acts as an antenna chromophore for the metal ions) having appropriate triplet state (T_1) energy in the scope of 22,000–27,000 cm^{-1}, which is essential for it [4, 12, 17, 25, 35, 36]. The effectiveness of energy transference and the kinetics of non-luminous transitions, that is, $^5D_0 \rightarrow T_1$ (Eu^{3+}, k_1^{Eu}) and $^5D_4 \rightarrow T_1$ (Tb^{3+}, k_1^{Tb}) are affected by the triplet state energy (T_1) of the organic linkers, which likewise influence the kinetics of the radiative transitions of the $^5D_0 \rightarrow ^7F_J$ transition of Eu^{3+} (k_{DF}^{Eu}) and of the $^5D_4 \rightarrow ^7F_J$ transition of Tb^{3+} (k_{DF}^{Tb}) [1, 4, 12]. On a fundamental level, one can adjust the temperature-dependent luminescence properties of the characteristic transitions of both the metal ions by utilizing various kinds of organic linkers with flexible T_1, and by this strategy, sensitivity of the M′Ln-MOF luminous thermosensors can be enhanced [3, 23]. If the triplet excited state (T_1) of the ligand is altogether higher than the accepting level $^5D_4$ (20,500 cm^{-1}) of Tb^{3+} and $^5D_0$ (17,200 cm^{-1}) of Eu^{3+}, at that point, the gap can ensure the viable sensitization and efficient energy transference from the organic ligand to both the metal ions at increased temperature based on the thermally determined phonon-assisted transference mechanism, and all the while restrict the energy back-move. Likewise, the energy transference from Tb^{3+} to Eu^{3+} in these frameworks can be promoted by the above features [36]. On the off chance that the T_1 of the ligand is not much higher, then, the energy gap between the two states, that is, T_1 of the ligand and accepting level $^5D_4$ of Tb^{3+} is less contrasted with the distinction between ligand's T_1 and accepting level $^5D_0$ of Eu^{3+}. It clearly implies that there must be back transference of energy from Tb^{3+} to ligand's T_1 at an increased temperature and the relating thermally determined quenches are much more grounded than that of Eu^{3+} ion. Consequently, the luminescence properties of the Eu^{3+} is increasingly effective by the sensitization of the organic ligands, which is additionally affirmed by the QY estimation [59]. The sensitivity of the mixed Ln-MOFs increases by the energy transference method from Tb^{3+} to Eu^{3+}. This energy transference process increases with the expansion in temperature because of the activation of nonradiative decay pathways, which facilitates more energy transference from the $^5D_4$ level of Tb^{3+} to $^5D_0$ level of Eu^{3+} ions.

2.3 Approaches to Measure Temperature

Luminescence thermometry envelops three primary methodologies to measure temperature:
1. Spectral shift of a given transition
2. Emission intensity estimations, utilizing the incorporated intensity of a single transition or a couple of transitions
3. Lifetime estimations, utilizing the decay outlines of the emitting energized states [6, 7, 20, 38, 45]

Basically, all MOF thermometers determine the absolute temperature through the estimation of the intensities of two transitions of distinct emitting centers. Most of the thermometers utilize the intensity proportion of the $^5D_4 \rightarrow {}^7F_5$ (Tb^{3+}) and the $^5D_0 \rightarrow {}^7F_2$ (Eu^{3+}) transitions as the thermometric parameter [8, 11]. The temperature reliance of the emission intensities is, mostly, administered by the thermally driven energy transference and back transference between the ligand and metal ion (Figure 2.2).

For dual center thermometers, the regularly utilized transformation of integrated intensity into temperature is made by means of the thermometric parameter, Δ:

$$\Delta = I_1 / I_2 \qquad (2.1)$$

where I_1 and I_2 are the integrated intensities of the two transitions [38].

2.4 Execution of Thermometers

The execution of luminescent thermometers is analyzed utilizing the accompanying parameters such as relative thermal sensitivity, temperature uncertainty, spatial and temporal resolution, and repeatability and reproducibility [38].

2.4.1 Relative Thermal Sensitivity

The relative thermal sensitivity S_r shows the relative alteration of the thermometric parameter per degree of temperature alteration:

$$S_r = \frac{1}{\Delta} \left| \frac{\partial \Delta}{\partial T} \right| \tag{2.2}$$

This parameter generally expressed in units of percentage (%) change per kelvin (K) of temperature change (%/K) [38, 43]. The maximum value of S_r is expressed as S_m. $S_r > 0$ (negative sensitivity is trivial):

$$S_a \text{(absolute sensitivity)} = \left| \frac{\partial \Delta}{\partial T} \right| \tag{2.3}$$

S_r has a significant favorable position of being free of the nature of the temperature sensors (i.e., mechanical, electrical and luminescent), permitting the immediate and quantitative correlation of various frameworks. Most of the Ln-MOF temperature sensors revealed so far recorded S_a change or without a doubt do not calculate the sensitivity.

2.4.2 Temperature Uncertainty (δT)

The temperature uncertainty (or temperature resolution) δT or smallest temperature alteration might be found out in a given estimation. Accepting that the temperature uncertainty of a temperature sensor results from changes in the thermometric parameters Δ, δT is given by the primary term of the extension of the temperature alteration with Δ [2, 38],

$$\delta T = \left| \frac{\partial T}{\partial \Delta} \right| \delta \Delta \tag{2.4}$$

where δT is the uncertainty in the assurance of Δ. Communicating condition (2.2) regarding S_r results in the following equation:

$$\delta T = \frac{1}{S_r} \frac{\delta \Delta}{\Delta} \tag{2.5}$$

where $\delta\Delta/\Delta$ is known as the relative uncertainty on Δ. Also, subsequently, ΔT relies upon the thermometer execution.

2.4.3 Spatial and Temporal Resolution

At the point when the temperature is estimated at various spatial positions, the spatial resolutions of the estimation (δx) are characterized as the base separation between points introducing a temperature contrast bigger than δT (or, proportionally, bigger than the temperature sensor's sensitivity). δx is evaluated utilizing the following expression [24]:

$$\delta x = \frac{\delta T}{|\nabla T|_{max}} \qquad (2.6)$$

where $||_{max}$ is the most extreme temperature inclination of the mapping. For a one-dimensional (1D) temperature outline, the gradient is $||_{max} = |\partial T/\partial x|_{max}$.

The temporal resolution of the assessment (∂t) is the base time span between assessments with temperature greater than ∂T:

$$\delta t = \frac{\delta T}{|\partial T/\partial t|_{max}} \qquad (2.7)$$

where $|\partial T/\partial x|_{max}$ is the most extreme temperature alteration per unit of time. The estimation of spatial–temporal resolution acquired by utilizing luminous temperature sensors has been looked at with those deliberate utilizing other (nonluminous) noncontact methods, for example, thermos-reflectance and Raman spectroscopy [5].

2.4.4 Repeatability and Reproducibility

Repeatability denotes the difference in repeat estimations made under alike circumstances. The repeatability of a temperature detector's readout upon temperature cycling is evaluated utilizing the following expression:

$$R = 1 - \frac{\max(|\Delta c - \Delta i|)}{\Delta} \qquad (2.8)$$

where Δc is the mean thermometric parameter and Δi is the value of each assessment of the thermometric parameter.

In this chapter, efforts have been made to summarize the recent reports based on Eu/Tb MOF luminous temperature sensors.

2.4.4.1 Eu$_{0.0069}$Tb$_{0.9931}$-DMBDC

The first Tb/Eu-mixed rare-earth MOF-based luminous temperature sensor Eu$_{0.0069}$Tb$_{0.9931}$-DMBDC was accounted by taking 2,5-dimethoxy-1,4-benzenedicarboxylate (DMBDC) as the organic linker [11]. Such type of Ln-MOF functions as a ratiometric thermometer based on the intensity proportion of two emission peaks, intensities of Tb^{3+} ($^5D_4 \rightarrow {}^7F_5$) at 545 nm and Eu^{3+} ($^5D_0 \rightarrow {}^7F_2$) at 613 nm. Single rare earth metal-based MOFs (Tb-DMBDC and Eu-DMBDC) were prepared by the solvothermal method and then by fluctuating the initial molar proportions of both Eu(NO$_3$)$_3$ to Tb(NO$_3$)$_3$, a bimetallic Eu^{3+}/Tb^{3+} mixed MOF (having common formula as (Eu$_x$Tb$_{1-x}$)$_2$(DMBDC)$_3$ (H$_2$O)$_4$·DMF·H$_2$O (Eu$_x$Tb$_{1-x}$-DMBDC), x = 0.0011, 0.0046 and 0.0069) is prepared by the similar synthetic method. Single-crystal X-ray diffraction (XRD) studies confirmed that all the synthesized MOFs are isostructural. Considering Tb-DMBDC as a representative, a 3D rod packing framework (Figure 2.3(a)) is formed by the association of each Tb^{3+}-carboxylate infinite secondary building unit (SBU) with other six chains by the phenylene part of the DMBDC organic ligands [11]. Temperature-dependent photoluminescence (PL) spectra of Tb$_x$Eu$_{1-x}$-DMBDC show both the characteristic emission peaks of Tb^{3+} and Eu^{3+}, concurrently. It is additionally worth to take note of that there is no ligand peak detected; it clearly indicates that DMBDC is an admirable antenna chromophore for sensitizing both trivalent Ln ions in MOF matrix. The excited triplet state energy (T$_1$) of DMBDC is 23,306 cm^{-1}, which is not much higher than that of the discharging level of Tb^{3+} ($^5D_4$, 20,500 cm^{-1}). An energy back exchange occurs from DMBDC to Tb^{3+} by decreasing the sensitivity. The energy gap between T$_1$ of DMBDC and radiating level of Eu^{3+} ($^5D_0$, 17,200 cm^{-1}) is higher than that of Tb^{3+} and due to no energy back transfer sensitization of Eu^{3+} by DMBDC is higher.

The temperature-dependent PL spectra of MOFs were inspected concerning both intensity and lifetime as parameters. At very low temperature of 10 K, both Eu^{3+} (at 613 nm) and Tb^{3+} (at 545 nm) having emission bands are of correspondence. At the point when temperature increases from 10 to 300 K, there is a decline in intensity of Tb^{3+} and that of Eu^{3+} increases, suggesting energy transference from Tb^{3+} to Eu^{3+}. It is also discovered that the lifetime of Tb^{3+} in Eu$_{0.0069}$Tb$_{0.9931}$-DMBDC is shorter than that of Tb-DMBDC, while that of Eu^{3+} is longer than Eu-DMBDC. This clearly indicates the energy gain in Eu^{3+} from Tb^{3+}. This Ln-MOF thermometer works in the temperature range of 10–300 K with relative sensitivity of 0.38%/K. Also, the color change from green to red (Figure 2.3(d)) with the temperature range of 10–300 K can be seen in the naked eye or camera.

2.4.4.2 Tb$_{0.9}$Eu$_{0.1}$PIA

Rao et al. [36] reported a bimetallic Ln-MOF temperature sensor Tb$_{0.9}$Eu$_{0.1}$PIA (PIA = 5-(pyridin-4-yl]isophthalate). The reaction between H$_2$PIA and Ln(NO$_3$)$_3$ by

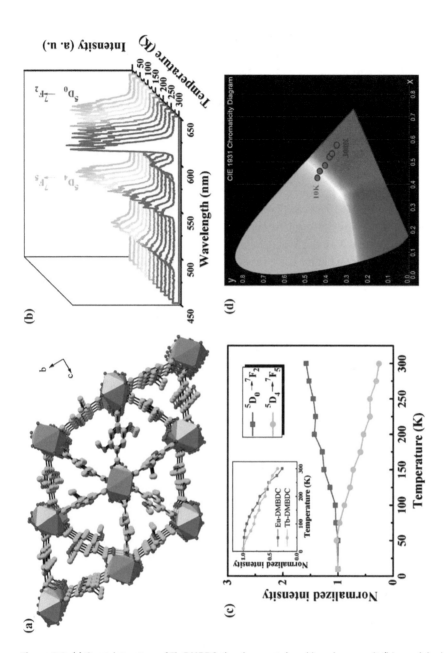

Figure 2.3: (a) Crystal structure of Tb-DMBDC showing crystal packing along a-axis (blue polyhedral, Tb; green, C; red, O). (b) Temperature-dependent PL spectra of $Eu_{0.0069}Tb_{0.9931}$-DMBDC (T = 10–300 K). (c) Temperature reliance of normalized intensities of the $^5D_4 \rightarrow {}^7F_2$ (534–562 nm) and $^5D_0 \rightarrow {}^7F_5$ (605–633 nm) transitions in $Eu_{0.0069}Tb_{0.9931}$-DMBDC and (inset) for Tb-DMBDC and Eu-DMBDC, respectively. (d) CIE chromaticity diagram for $Eu_{0.0069}Tb_{0.9931}$-DMBDC showing luminescence color at various temperatures. Reprinted with permission from [11]. Copyright (2012) American Chemical Society.

the hydrothermal method gives rise to the Tb/Eu mixed bimetallic MOFs $Tb_{1-x}Eu_xPIA$ (x = 0.01, 0.05, 0.10, 0.20, 0.50 and 0.80). M'Ln-MOF has a sensitivity of 3.53%, which is around 9 times greater when compared to the previously made $Eu_{0.0069}Tb_{0.9931}$-DMBDC having sensitivity of 0.38% [11]. Isostructural nature of all the as-synthesized LnPIA MOFs is confirmed by powder crystal XRD. From single-crystal XRD, it is found that TbPIA takes shape in the orthorhombic system with space group Pba_2, and the asymmetric unit consists of a single Tb atom, two PIA organic linkers and 2.5 molecules of water [36]. Four PIA linkers give seven O atoms from their carboxylate units, and the two coordinated H_2O molecules donate the rest two O atoms to form the nine coordinated Tb^{3+} ion. A 2D-layered structure of TbPIA (Figure 2.4(b)) is formed when the mononuclear $Tb(COO)_4$ SBUs (Figure 2.4(a)) are crossed over by means of PIA ligands. Very dense 3D framework is shaped by combining the 2D layers with the help of weak intermolecular Van der Waals forces and aromatic π–π collaborations. Upon excitation at 330 nm, the ligand H_2PIA displays a weak emission at 408 nm, because of π → π* transitions, while Tb and Eu show their characteristic emission peaks separately (λ_{exc} = 360 nm). In the temperature-dependent PL spectra, the luminescent conduct of Tb^{3+} in TbPIA is practically identical to that of TbDMBDC, while that of Eu^{3+} in EuPIA declines substantially more gradually than those of EuDMBDC with the expansion of temperature from 14 to 300 K. The predictable energy difference of 8,666 cm^{-1} between the T_1 state H_2PIA (25,866 cm^{-1}) and the Eu^{3+} radiating level ($^5D_0$, 17,200 cm^{-1}) in EuPIA is huge than the distinction of 6,106 cm^{-1} in EuDMBDC to deny thermally determined elimination. The temperature-dependent PL spectra of the mixed Ln-MOF $Tb_{0.9}Eu_{0.1}PIA$ are shown in Figure 2.4(c). Predictably, the representative mixed rare earth metal-based MOF $Tb_{0.9}Eu_{0.1}PIA$ displays an altogether extraordinary temperature-dependent luminescent behavior concerning the emissions of both Eu^{3+} (615 nm) and Tb^{3+} (615 nm). In the low-temperature region when expansion in the temperature range occurs from 14 to 100 K, the emission intensity of both the ions, that is, Tb^{3+} and Eu^{3+} remains unaltered.

Effective energy transference from Tb^{3+} to Eu^{3+} occurs with the temperature extension of 100–300 K. As a result of this phenomenon, the emission intensity of Tb^{3+} diminished, while that of Eu^{3+} increases. This additionally results in the lifetime of $^5D_4$ in $Tb_{0.9}Eu_{0.1}PIA$ diminished step by step as the increment of temperature occurs, while that of $^5D_0$ increases from 374.66 to 433.83 μs with temperature varying from 14 to 225 K and afterward diminished to 378.01 μs at 300 K as shown in Figure 2.4(d). The sensitivity can be additionally improved to be around 20% and 6% for $Tb_{0.99}Eu_{0.01}PIA$ and $Tb_{0.95}Eu_{0.05}PIA$ by expanding the proportion of Tb^{3+} in the bimetallic Ln-MOFs.

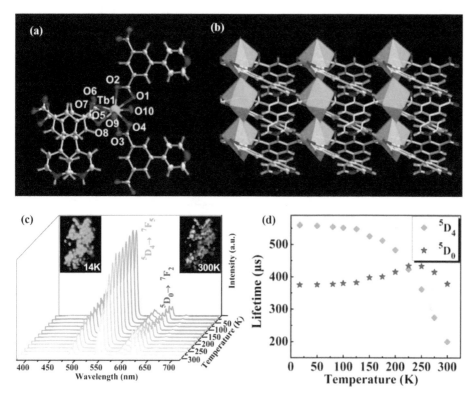

Figure 2.4: (a) Coordination environment of rare earth ion. (b) Crystal structure of TbPIA showing 2D structure. (c) PL spectra of $Tb_{0.9}Eu_{0.1}PIA$ (temperature range = 14–300 K). (d) Temperature-dependent lifetime of $^5D_4$ (at 546 nm) and $^5D_0$ (at 615 nm) for $Tb_{0.9}Eu_{0.1}PIA$ (10–300 K) under excitation at 360 nm. Reprinted with permission from [36]. Copyright (2013) American Chemical Society.

2.4.4.3 $Tb_{0.957}Eu_{0.043}$cpda

A $Tb_{0.957}Eu_{0.043}$cpda-based MOF was reported to identify and visualize temperature over an extensive range from 40 to 300 K [15]. Tbcpda, Eucpda and $Tb_{0.957}Eu_{0.043}$cpda were synthesized by the solvothermal method. From, single-crystal XRD and powder XRD (P-XRD) analysis, it is found that all the synthesized MOFs are isostructural. From single-crystal XRD, it also known that Tbcpda solidifies in the monoclinic structure (space group C2/c) (Figure 2.5(a)). Tb^{3+} ion forms eight-coordinated polyhedra by linking with six O atoms and two N atoms. Out of them, six are contributed by two tridentate chelated ONO atoms of two H_3cpda linkers, the nearby H_3cpda ligand contributes one O atom and the other O atom is contributed by the coordinated water molecule. The 1D zigzag chains are formed by the association of neighboring Tb^{3+} with the help of monodentate carboxyl units of the H_3cpda linker along the b-axis, and afterward by π–π contacts, the 1D chains are packed to frame the 3D system structure.

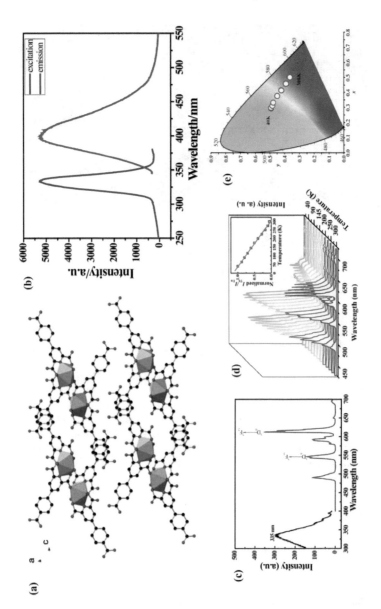

Figure 2.5: (a) View of crystal packing along the *b*-axis (blue polyhedral, Tb; black, C; red, O; blue, N). Room-temperature excitation and emission spectra of (b) H_3cpda and (c) $Tb_{0.957}Eu_{0.043}$cpda. (d) Temperature-dependent emission spectra of $Tb_{0.957}Eu_{0.043}$cpda under excitation at 335 nm (temperature range = 40–300 K); the inset shows the intensity proportion curve of Tb^{3+} (546 nm) and Eu^{3+} (615 nm) for $Tb_{0.957}Eu_{0.043}$cpda. (e) CIE chromaticity diagram of $Tb_{0.957}Eu_{0.043}$cpda displaying temperature-dependent luminescence color changing property [15]. Reproduced with permission from the Royal Society of Chemistry.

Room-temperature emission spectra of the ligand is found at 410 nm (λ_{exc} = 330 nm) as shown in Figure 2.5(b). Similarly, with excitation at 330 nm Tb-cpda and Eu-cpda, both display the characteristic emission peaks of Tb^{3+} and Eu^{3+}, respectively. $Tb_{0.957}Eu_{0.043}$cpda contains both the characteristic peaks of Tb^{3+} and Eu^{3+} as depicted in Figure 2.5(c), but no ligand peak is observed, which implies that the ligand acts as a sensitizer, and not as an emissive center. Temperature-dependent PL spectra of $Tb_{0.957}Eu_{0.043}$cpda (Figure 2.5(d)) show that with the expansion in temperature from 40 to 300 K, there is decline in the intensity of Tb^{3+}, while that of Eu^{3+} increases. This type of luminescence behavior is totally different from Tbcpda and Eucpda, which shows not many changes over the entire temperature range and is not temperature delicate. From these discussions, it is observed that the energy transference procedure from Tb^{3+} to Eu^{3+} is weak at cryogenic temperature, while strongly increased at high-temperature region due to the thermal activation of non-radiative decay pathways. In this case, the triplet excited state of the ligand is located at 27,027 cm^{-1} which is a lot higher than that of the emitted level of Tb^{3+} ($^5D_4$) and Eu^{3+} ($^5D_0$), which prevents the back energy transfer from luminescent ions to organic ligand (cpda). This results in terms of increase in sensitivity of the ratiometric luminescent thermometer. As temperature increases from 40 to 300 K, there is color change from green to red and is shown in CIE chromaticity diagram (Figure 2.5(e)).

In Figure 2.6(a), it is shown that the lifetime of $^5D_4$ decreases to 90% as temperature increases from 15 to 300 K, while that of $^5D_0$ does not show any critical lessening just because of energy move from Tb^{3+} to Eu^{3+}. The effectiveness of energy transference (as shown in Figure 2.6(b)) is higher as compared to $Eu_{0.0069}Tb_{0.9931}$-DMBDC [11].

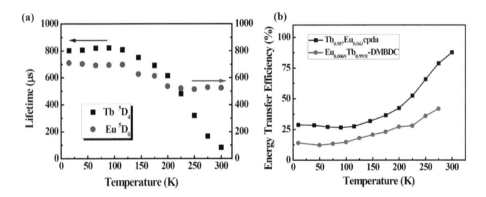

Figure 2.6: (a) $Tb_{0.957}Eu_{0.043}$cpda showing temperature-dependent lifetime of $^5D_4$ and $^5D_0$ (15–300 K observed at 546 and 615 nm, correspondingly (λ_{exc} = 335 nm)). (b) Efficiency of energy transference from Tb^{3+} to Eu^{3+} in $Tb_{0.957}Eu_{0.043}$cpda and $Tb_{0.957}Eu_{0.043}$cpda based on temperature [15]. Reproduced with permission from the Royal Society of Chemistry.

2.4.4.4 Eu$_{0.0618}$Tb$_{0.9382}$L

A mixed lanthanide coordination polymer (CP) [(Eu$_{0.0616}$Tb$_{0.9382}$)(L)$_2$(NO$_3$)$_2$]·Cl·2H$_2$O commonly known as Eu$_{0.0618}$Tb$_{0.9382}$L (L = 1,4-bis(pyridinil-4-carboxylato)-1,4-dimethylbenzene) was developed by doping Eu^{3+} ion into an isostructural Tb^{3+}-CP, which can act as a self-referenced luminescent temperature detector by using intensity proportion of two emissions, that is, Tb^{3+} (544 nm) and Eu^{3+} (613 nm) are more efficient than those compounds exhibiting single emission (EuL/TbL) [32]. This mixed Ln-CP is synthesized by the solvothermal reaction of [H$_2$L]Cl$_2$ with Ln(NO$_3$)$_3$·6H$_2$O in DMF as solvent, heated for 74 h at 85 °C, and colorless crystals of Ln-CPs having tetragonal shape are formed. As P-XRD reveals that all the Ln-CP and mixed Ln-CP are isostructural, EuL is taken as the representative to describe their structure. EuL crystallizes in the tetragonal structure having space group I4/m, where the Eu ion is placed at the center adopting a distorted dodecahedron geometry, coordinated by eight oxygen atoms (four from 4L linkers and rest two from nitrate ions). The geometry of Eu and packing diagram of EuL are shown in Figure 2.7(a, b). The free ligand [H$_2$L]Cl$_2$ shows an expansive band with a most extreme at 397 nm (λ_{exc} = 203 nm) due to π→π* electronic transitions (as shown in Figure 2.7(c)). In case of EuL and TbL, the characteristic emission peaks of Eu^{3+} and Tb^{3+} are shown by exciting at 352 and 357 nm, respectively. In the M'Ln-CPs, Eu$_{0.0311}$Tb$_{0.9689}$L and Eu$_{0.0618}$Tb$_{0.9382}$L, the blue light emission from the organic ligand disappeared but the characteristic emission peaks of $^5D_0 \rightarrow {^7F_{1-4}}$ (Eu^{3+}) and $^5D_4 \rightarrow {^7F_{6-2}}$ (Tb^{3+}) transitions were appeared, suggesting [H$_2$L]Cl$_2$ linker is an outstanding (antenna) chromophore for sensitization of both Eu^{3+} and Tb^{3+} ions. Figure 2.7(d) displays the normalized temperature-dependent PL spectra of Eu$_{0.0618}$Tb$_{0.9382}$L. As temperature rises from 25 to 200 K, the luminescence intensity of Tb^{3+} declines and that of Eu^{3+} rises representing effective energy transference from Tb^{3+} to Eu^{3+} ion occurs.

2.4.4.5 [Eu$_{0.7}$Tb$_{0.3}$(D-cam)(Himdc)$_2$(H$_2$O)$_2$]$_3$

In 2014, a new bifunctional rare-earth-based luminescent carboxylate framework [Eu$_{0.7}$Tb$_{0.3}$(D-cam)(Himdc)$_2$(H$_2$O)$_2$]$_3$ was reported [19]. This Ln-MOF is a reasonable test for detecting temperature in a more extensive range from 100 to 450 K having a maximum sensitivity of 0.11%/K at high temperature of 450 K. In addition, it is exceptionally steady in an extensive pH range and a few boiling solvents. It is having pH detecting capacity over an extensive pH run from 2 to 11 and displays exceptionally delicate and straight pH response in the physiological range (pH = 6.8–8.0). Single metal-based Ln-MOFs, [Ln$_2$(D-cam)(Himdc)$_2$(H$_2$O)$_2$] (Ln = Eu, Tb) and bimetallic mixed Ln-MOFs [Eu$_{0.7}$Tb$_{0.3}$(D-cam)(Himdc)$_2$(H$_2$O)$_2$]$_3$ were prepared by the hydrothermal reaction of Ln^{3+} ions and mixed Tb^{3+}/Eu^{3+} salts with D-camphoric acid (D-H$_2$cam) and 4,5-imidazole dicarboxylic acid (H$_3$imdc) in a solvent having mixture

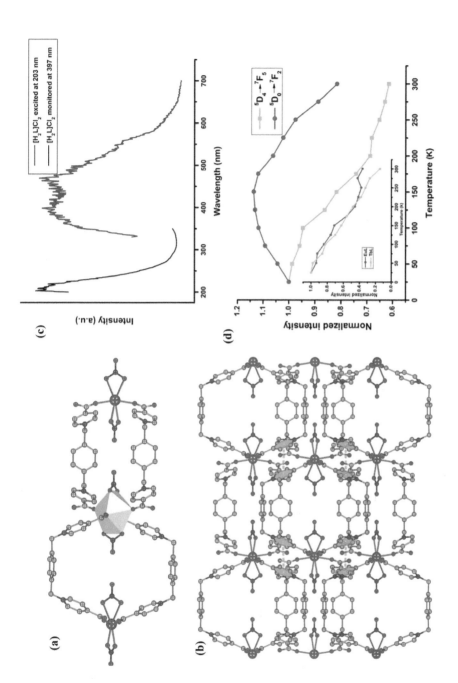

Figure 2.7: (a) A distorted dodecahedron geometry of europium ion. (b) The packing diagram of EuL along *b*-axis (green dot line: π . . . π contact). (c) RT excitation (black) and PL (blue) spectra of ([H_2L]Cl_2) in the solid state. (d) Temperature-dependent normalized emission intensity of the $^5D_4 \rightarrow {}^7F_5$ and $^5D_0 \rightarrow {}^7F_2$ transition for $Eu_{0.0618}Tb_{0.9382}L$; the inset shows those for TbL and EuL [32]. Reproduced with permission from the Royal Society of Chemistry.

of DMF and water in it and are heated at 180 °C for 48 h, forming colorless crystals of Ln-MOF/M′Ln-MOF (M′ = mixed Ln-MOF). Single-crystal XRD confirmed that all the as-synthesized MOFs are isostructural and take shape in the orthorhombic structure having space group Pna2$_1$. Two trivalent Ln ions, single D-cam^{2-} linker, two Himdc^{2-} organic linkers and two molecules of coordinated H$_2$O are coordinated to form the asymmetric unit. Accepting the Eu-MOF, for instance, every one of the two Eu^{3+} ions is in distorted hendecahedral geometry.

The crystal structure of Eu-MOF and the temperature-dependent PL spectra of [Eu$_{0.7}$Tb$_{0.3}$(D-cam)(Himdc)$_2$(H$_2$O)$_2$] are shown in Figure 2.8. As temperature increases from 100 to 450 K, the intensity of Eu^{3+} at 613 nm becomes dominating over the intensity of Tb^{3+} at 545 nm. In addition to that, the lifetime of ^{5}D$_4$ of Tb^{3+} decreases more in comparison to their pure MOFs, while that of Eu^{3+} (^{5}D$_0$) shows no significant decrease with the increments in temperature, indicating easy energy transference from Tb^{3+} to Eu^{3+}. The effectiveness of energy transference increases with the increase in temperature.

2.4.4.6 Tb$_{0.95}$Eu$_{0.05}$HL

Mixed-metal Ln-MOF Tb$_{0.95}$Eu$_{0.05}$HL with good sensitivity over an extensive temperature range from 4 to 290 K was reported [29]. Synthesis of Ln-MOFs was executed by hydrothermal reactions of 5-hydroxy-1,2,4-benzenetricarboxylic acid (H$_4$L) with LnCl$_3$·6H$_2$O.

P-XRD and infrared (IR) analysis confirmed that all these frameworks are isomorphous and take shape in the monoclinic structure with space group C$_{2/c}$. The asymmetric unit of TbHL is made up of a Tb^{3+} ion, one ligand of HL^{3-} and 1 1/2 molecules of H$_2$O. A distorted triaugmented triangular prism geometry is formed by Tb^{3+} ion, which is coordinated by nine oxygen atoms, out of them seven oxygen givers originate from five carboxylate units and the water molecules contribute the other two O atoms. The bond distances between the Tb and O atom lie in the range of 2.26–2.64 Å. The repeating unit of [Tb$_2$(COO)$_6$(H$_2$O)$_3$] is formed by linking the nearby Tb^{3+} ions with the help of two crossing over O atoms from the carboxyl groups, which are then connected by a water molecule to create the SBU. The nearby Tb^{3+} to Tb^{3+} separations are 4.212 and 4.532 Å. The 3D framework is made by connecting SBUs with four nearby SBUs by means of the linker HL^{3-} (Figure 2.9(b)). TbHL shows characteristic emission peaks of the Tb^{3+} ion (λ_{exc} = 325 nm) and EuHL displays the characteristic peak of Eu^{3+} ion, whereas the doped compound Tb$_{0.95}$Eu$_{0.05}$HL shows both the Tb^{3+} and Eu^{3+} ion characteristic peaks simultaneously (Figure 2.9(c, d)). In this case, the energy difference between T$_1$ (26,600 cm^{-1}) of the organic linker and radiating state of Tb^{3+} (^{5}D$_4$, 20,500 cm^{-1}) is high (around 6,100 cm^{-1}), which infers that the energy transference from T$_1$ to Tb^{3+} happens without energy back transfer. When temperature increment occurs from 4 to 290 K, there is a decrease in intensity of Tb^{3+}, while that of Eu^{3+} rises significantly as shown in the temperature-dependent PL emission

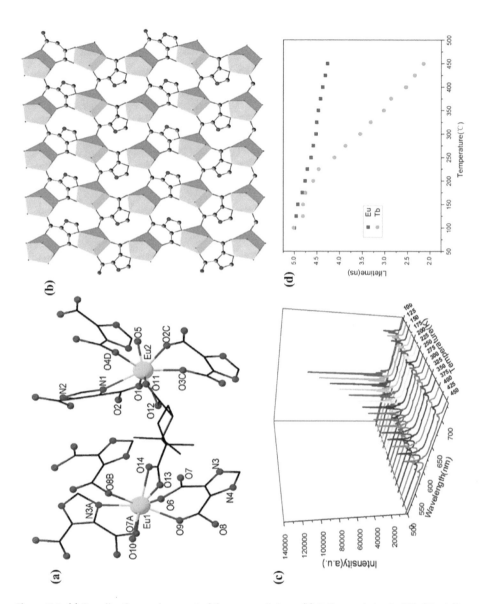

Figure 2.8: (a) Coordination environment of the rare earth ions. (b) Unilayered sheet of {Eu(Himdc)}$_3$ along the c-axis. (c) The luminescent emissions of [Eu$_{0.7}$Tb$_{0.3}$(D-cam)(Himdc)$_2$(H$_2$O)$_2$] at different temperatures. (d) Temperature-dependent lifetime of [Eu$_{0.7}$Tb$_{0.3}$(D-cam)(Himdc)$_2$(H$_2$O)$_2$] (545/613 nm) (100–450 K). The decay curves are observed at 545/613 nm [λ_{exc} = 277 nm) [19]. Reproduced with permission from the Royal Society of Chemistry.

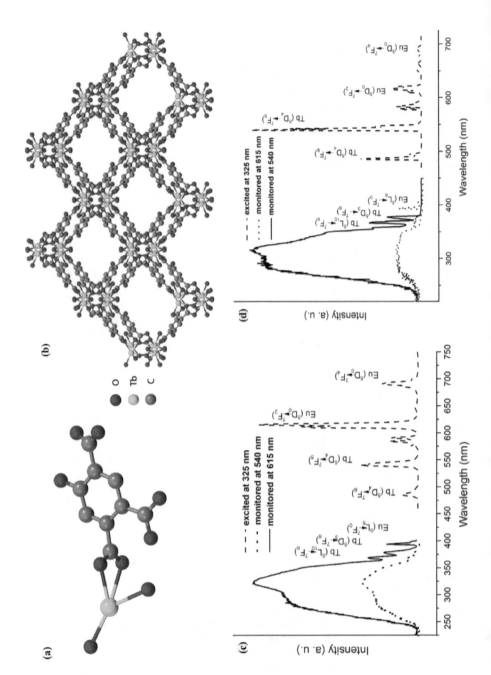

Figure 2.9: (a) Asymmetric unit of a rare earth ion. (b) Framework of TbHL along the *c*-axis. Excitation and emission spectra of Tb$_{0.95}$Eu$_{0.05}$HL at (c) RT and (d) 4 K. Reprinted with permission from [29]. Copyright (2015) American Chemical Society.

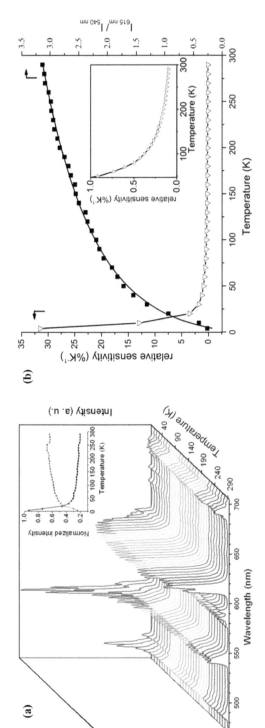

Figure 2.10: (a) Solid-state PL spectra of $Tb_{0.95}Eu_{0.05}HL$ under excitation at 325 nm (temperature range of 4–290 K); the inset shows normalized intensity of $^5D_4 \rightarrow {}^7F_5$ (Tb^{3+}-black, 540 nm) and $^5D_0 \rightarrow {}^7F_2$ (Eu^{3+}-red, 615 nm) in $Tb_{0.95}Eu_{0.05}HL$ under excitation at 325 nm. (b) Temperature-dependent emission intensity proportion curve (right axis, black line) and relative sensitivity curve (left axis); the inset shows the enlarged relative sensitivity curve covering the temperature range from 50 to 300 K. Reprinted with permission from [29]. Copyright (2015) American Chemical Society.

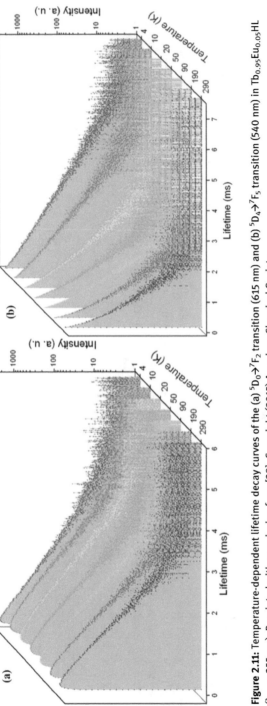

Figure 2.11: Temperature-dependent lifetime decay curves of the (a) $^5D_0 \to {}^7F_2$ transition (615 nm) and (b) $^5D_4 \to {}^7F_5$ transition (540 nm) in $Tb_{0.95}Eu_{0.05}HL$ (λ_{exc} = 355 nm). Reprinted with permission from [29]. Copyright (2015) American Chemical Society.

spectra (Figure 2.10(a)). This is nothing but due to the phonon helped energy move from Tb^{3+} to Eu^{3+} and phonon helped energy relocation among neighboring Tb^{3+} ions. This Ln-MOF shows high affectability over the temperatures extend from 4 to 290 K with most extreme relative sensitivity of 31% at 4 K.

There is a dramatic decrease in the lifetime of Tb^{3+} emission from 4 to 290 K, while that of Eu^{3+} decreases slightly (Figure 2.11). This result implies the increase in possibility of energy transference from Tb^{3+} to Eu^{3+} as the temperature increases.

2.4.4.7 $Tb_{0.98}Eu_{0.02}$-OA-DSTP and $Tb_{0.98}Eu_{0.02}$-BDC-DSTP

Two sorts of exceptionally steady and delicate rare earth metal-based temperature sensors $[Tb_{1-x}Eu_x(OA)_{0.5}(DSTP)]\cdot 3H_2O$ and $[Tb_{1-x}Eu_x(BDC)_{0.5}(DSTP)]\cdot 2H_2O$ (x = 0.01, 0.02) were prepared with the help of hydrothermal method, by taking 2,4-(2,2′:6′,2′-terpyridin-4′-yl)-benzenedisulfonic acid (H_2DSTP) as the primary ligand and altering the secondary ligand with oxalic acid (OA) or 1,4-benzene dicarboxylic acid (BDC) [47]. Noticeable change in the ratiometric temperature detecting properties of the mixed Ln-MOF temperature sensors can be seen due to the addition of the supplementary ligands BDC and OA. From single-crystal crystallography, it is found that all Ln-MOFs were isostructural. In the two types of Ln-MOFs, DSTP takes up the similar coordination mode. A 2D-layer Ln-DSTP was formed by connecting four lanthanide ions with the help of three chelating N atoms of terpyridine, bidentate bridging and monodentate sulfonic units. An ultimate 2D polymeric structure of Ln-OA-DSTP (Figure 2.12(a)) was formed by connecting two metal centers utilizing two carboxylate groups of the organic linker OA and then embedded into the 2D Ln-DSTP layer, while the ligand BDC interface four metal centers with the help of two bidentate carboxylate units and connects the 2D-layered Ln-DSTP, giving the resulting 3D polymeric structure of Ln-BDC-DSTP (Figure 2.12(b)).

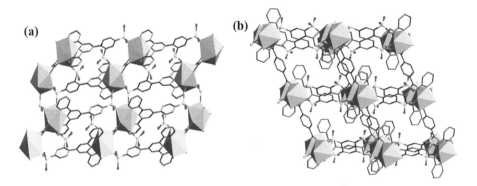

Figure 2.12: (a) Two-dimensional polymeric structure of Ln-OA-DSTP and (b) 3D polymeric framework of Ln-BDC-DSTP [47]. Reproduced with permission from the Royal Society of Chemistry.

An eight-coordinated dodecahedron geometry is adopted by the metal centers in Ln-OA-DSTP and Ln-BDC-DSTP without any coordinated H_2O molecules. The O and N atoms of the primary ligand completely occupy the structure, and a tight-holding framework is formed by the two secondary linkers which are required to improve the luminescence productivity and thermal strength of the mixed bimetallic M'Ln-MOF. The PL emission of $Tb_{0.98}Eu_{0.02}$-OA-DSTP is depicted in Figure 2.13. In the temperature-dependent PL study, the emission intensity of Tb^{3+} diminished significantly with the expanding temperature, while Eu^{3+} emission nearly remains unaffected for $Tb_{0.98}Eu_{0.02}$-OA-DSTP and it increments progressively for $Tb_{0.98}Eu_{0.02}$-BDC-DSTP [47]. In the temperature range of 77–225 K, there is overlapping between the emissive transitions of Eu^{3+} ($^5D_0\rightarrow{^7F_2}$) with Tb^{3+} ($^5D_4\rightarrow{^7F_3}$). That is why $^5D_0\rightarrow{^7F_2}$ transition cannot be preferred as a parameter to examine the luminescence properties of Eu^{3+}. As there is no overlay of the $^5D_0\rightarrow{^7F_4}$ transition with any transition of Tb^{3+}, its proportion with the characteristic transition ($^5D_4\rightarrow{^7F_5}$) of Tb^{3+} was taken as the most important parameter. The initial triplet state energy of DSTP is 19,050 cm^{-1}, which is marginally greater compared to the accepting level of Eu^{3+} ($^5D_0$, 17,200 cm^{-1}) and lesser than that of Tb^{3+} ($^5D_4$, 20,400 cm^{-1}). The energy transference occurs from T_1 of secondary ligands (i.e., OA and BDC) to the $^5D_4$ level of Tb^{3+} ion. BDC is having lowest triplet state than that of OA, which results in progressively effective energy back transference and also allows the transference of energy to Eu^{3+} from Tb^{3+}. Due to this energy transfer, the lifetime of $^5D_0$ (Tb^{3+}) in the bimetallic Tb/Eu mixed MOF becomes lower than pure MOF while that of Eu^{3+} emission increases.

This ratiometric thermometer works in the temperature range of 77–400 K. $Tb_{0.98}Eu_{0.02}$-BDC-DSTP shows maximum sensitivity of 2.8%/K at 225 K and $Tb_{0.99}Eu_{0.01}$- BDC-DSTP shows 3.9%/K at 200 K, which are greater than that of $Tb_{0.98}Eu_{0.02}$-OA-DSTP.

2.4.4.8 Ad/$Tb_{0.999}Eu_{0.001}$/BPDC

By taking adenine (Ad) and biphenyl-4,4'-dicarboxylic acid (BPDC) as two kinds of organic ligands, a mixed lanthanide complex, namely, Ad/$Tb_{0.999}Eu_{0.001}$/BPDC was designed and prepared by the hydrothermal reaction [40]. The compound acts as an efficient temperature detector in the range of 100–300 K as compared to $Tb_{0.999}Eu_{0.001}$/BPDC (containing a single ligand). To compare the temperature sensing ability of the mixed lanthanide complex, two types of single rare earth metal-based compounds Ad/Eu/BPDC and Ad/Tb/BPDC were additionally arranged. It is reported that both the pure and mixed lanthanide materials are isostructural, which is confirmed from XRD pattern. Both Ad/Eu/BPDC and Ad/Tb/BPDC show the characteristic emission peaks and colors of Eu^{3+} (red) and Tb^{3+} (green), respectively (λ_{exc} = 321 nm) (Figure 2.14(a)). Ad/$Tb_{0.999}Eu_{0.001}$/BPDC shows simultaneously both the characteristic peaks of Tb^{3+} ($^5D_4 \rightarrow {^7F_J}, J = 6-3$) and Eu^{3+} ($^5D_0 \rightarrow {^7F_J}, J = 0-4$). The CIE chromaticity diagram shows red color.

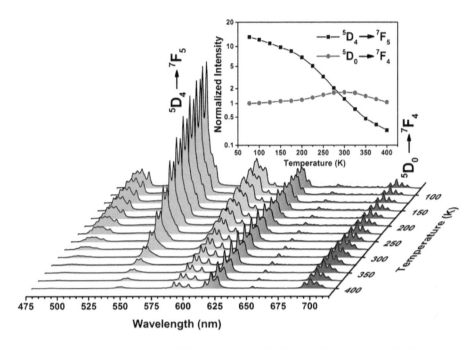

Figure 2.13: Solid-state PL spectra of $Tb_{0.98}Eu_{0.02}$-OA-DSTP (λ_{exc} = 360 nm, T = 77–400 K); the inset shows the normalized emission intensities of Tb^{3+} ($^5D_4 \rightarrow {}^7F_5$, 546 nm) and Eu^{3+} ($^5D_0 \rightarrow {}^7F_4$, 696 nm) [47]. Reproduced with permission from the Royal Society of Chemistry.

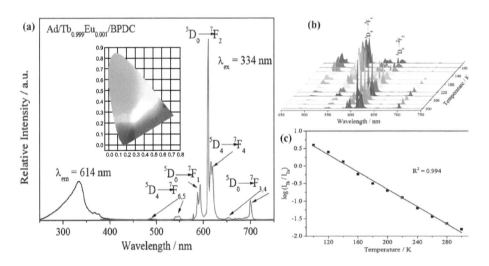

Figure 2.14: (a) Excitation (λ_{em} = 614 nm) and emission (λ_{exc} = 334 nm) spectra of Ad/$Tb_{0.999}Eu_{0.001}$/BPDC (inset: CIE chromaticity diagram indicating color in red region). (b) Temperature-dependent PL spectra of Ad/$Tb_{0.999}Eu_{0.001}$/BPDC (80–300 K). (c) Fitted curves for Ad/$Tb_{0.999}Eu_{0.001}$/BPDC. Reprinted with permission from [40].

With the expansion in temperature, there is noticeable decline in the luminescence intensity of Tb^{3+} at 545 nm, while that of Eu^{3+} increases significantly at 612 nm (Figure 2.14(b)), known from the temperature-dependent PL study. These two intensities were used as parameters in ratiometric thermometers. At 280 K, the intensity of Eu^{3+} emission is 5 times greater as compared to that at 100 K, while that of Tb^{3+} emission displays a slight decline in intensity to 4.62% at 100 K. This improved intensity of Eu^{3+} emission is nothing but due to the energy move from Tb^{3+} to Eu^{3+}, likewise the PL lifetime supports the same, that is, decrease in lifetime of Tb^{3+} from 148.7 to 21.0 µs, while that of Eu^{3+} does not show any significant decrease (1,539–1,445 µs). However, the Eu^{3+} luminescence decreases to a small extent at 300 K, due to the more intense molecular vibration. Figure 2.14(c) displays the dependence of $\log(I_{Tb}/I_{Eu})$ on temperature, which shows a decent direct relationship inside the temperature range of 100–300 K. But, lanthanide complex having only one ligand ($Tb_{0.999}Eu_{0.001}$/BPDC) shows a decent direct relationship within the temperature range of 150–250 K. Hence, it is concluded that introduction of second organic ligand into the complex increases the sensitivity of the Ln complex thermometer.

2.4.4.9 $Tb_{0.80}Eu_{0.20}$BPDA

$Tb_{0.80}Eu_{0.20}$BPDA was reported by Zhao et al. [60], which is having 7.5 times higher sensitivity as compared to $Tb_{0.99}Eu_{0.01}(BDC)_{1.5}(H_2O)_2$ [8]. Zhao et al. structured a ligand, to be specific, biphenyl-3,5-dicarboxylic acid (H_2BPDA), with a most minimal triplet state energy of 25,269 cm^{-1}, which can successfully sensitize both the trivalent rare earth metal ions (Tb^{3+} and Eu^{3+}). Energy-level distribution of the organic linker is firmly identified with the energy transference method in the framework. A drab little block-like crystal of LnBPDAs and $Tb_{0.80}Eu_{0.20}$BPDA was synthesized by the solvothermal strategy. Synthesis of the organic ligand is shown in Scheme 2.1. P-XRD concludes the isomorphous structure of all the prepared LnBPDA and $Tb_{0.80}Eu_{0.20}$BPDA, and the single-crystal XRD affirmed the monoclinic structure of the EuBPDA unit with space group P21/c and each unit comprises a single Eu molecule, a BPDA ligand, three molecules of DMF solvent and one NO^{3-} ion. The crystal structure of EuBPDA is shown in Figure 2.15(a, b), in which Eu^{3+} is nine-coordinated with O atoms, and the local geometry can be recognized as a distorted tricapped trigonal prism. The temperature-dependent PL of as-prepared LnBPDA and $Tb_{1-x}Eu_x$BPDA was studied. As shown in Figure 2.15(c), all of the mixes with various proportions of trivalent Ln ions (Tb^{3+} and Eu^{3+}) displayed emission spectra with various intensities for Tb^{3+} (544 nm) and Eu^{3+} (614 nm). Most often, the intensity proportion (Tb/Eu) is utilized as the thermometric parameter (since it has the particular favorable position of being free of the detector fixations in the medium and being a self-adjusting estimation of the temperature from the outflow spectra) to make sense of its temperature detection, as an appropriate luminescent intensity proportion somewhere in the range of Tb^{3+} and Eu^{3+} is essential. It is found that $Tb_{0.8}Eu_{0.2}BPDA$ is the best example to be inspected in detail, because $Tb_{0.9}Eu_{0.1}BPDA$ shows very low intensity at 614 nm and that of $Tb_{0.5}Eu_{0.5}BPDA$ is not predictable at 544 nm.

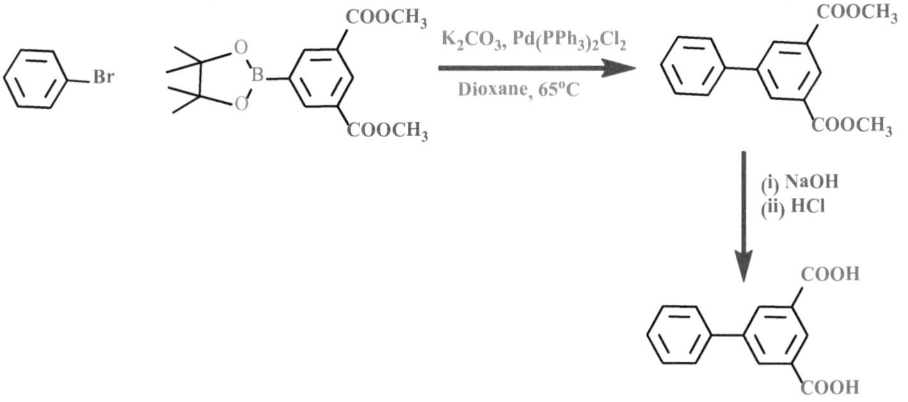

Scheme 2.1: Synthesis of H_2BPDA.

The sensitivity of $Tb_{0.8}Eu_{0.2}BPDA$ is situated in the physiological temperature range, which is totally unique in relation to different past materials showing good result in the temperature range from 10 to 300 K. The PL spectra are shown in Figure 2.15(d), which demonstrates that $Tb_{0.8}Eu_{0.2}BPDA$ is an admirable luminous temperature sensor in the range of 303–328 K. Surprisingly, the characteristic intensities of Tb^{3+} (544 nm, $^5D_4 \rightarrow {}^7F_5$) and Eu^{3+} (614 nm, $^5D_0 \rightarrow {}^7F_2$) for $Tb_{0.8}Eu_{0.2}BPDA$ show an interesting behavior. In other words, the intensities of the relating peaks are truly steady, regardless when there are increments in temperature from 10 to 300 K. In the temperature range of 10–300 K, $Tb_{0.8}Eu_{0.2}BPDA$ shows anomalous PL properties when contrasted with different examples 2′-fluoro-[1,1′:4′,1″-terphenyl]-3,3″,5,5″-tetracarboxylic acid can be credited to continued extraordinarily energy transference from the linker to the Tb^{3+} and Eu^{3+} ions. The refinement in $Tb_{0.8}Eu_{0.2}BPDA$ is not influenced by the correct energy levels and differences among the singlet and triplet energized states of the H_2BPDA linker, nonradiative decay and energy back exchange. This is a requirement for the incredible execution of $Tb_{0.8}Eu_{0.2}BPDA$ above RT, undoubtedly, in the extent of physiological temperatures. In addition, the lifetime of $Tb_{0.8}Eu_{0.2}BPDA$ is shorter than that of TbBPDA ($^5D_4$, Tb^{3+}), yet a more drawn out lifetime than EuBPDA ($^5D_0$, Eu^{3+}) from 293 to 328 K, with the color changing from yellow to red-yellow.

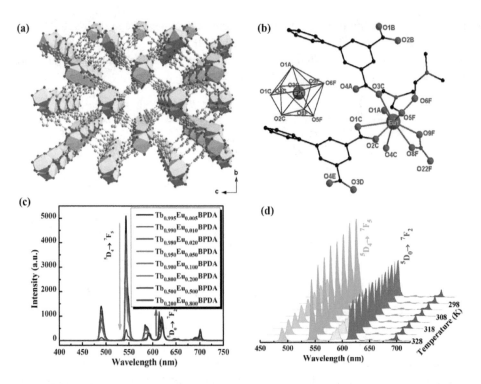

Figure 2.15: (a) Three-dimensional framework. (b) Coordination environment of rare earth ions (inset shows coordination polyhedron of Eu^{3+} ion). (c) Room-temperature emission spectra of $Tb_{1-x}Eu_xBPDA$ (λ_{exc} = 323 nm). (d) Temperature-dependent PL spectra of $Tb_{0.8}Eu_{0.2}BPDA$ under excitation at 323 nm (293–328 nm). Reprinted with permission from [56]. Copyright (2015) American Chemical Society.

2.4.4.10 $Eu_{0.0878}Tb_{0.9122}L$

In the year 2016, a new ratiometric luminous thermosensor $Eu_{0.0878}Tb_{0.9122}L$ was reported, and it shows a decent straight relationship with temperature in the range varying from 75 to 250 K [48]. In addition, white light emission was seen by shifting the λ_{exc} of $Eu_{0.0748}Tb_{0.9252}L$ compound. By taking 5-(4-tetrazol-5-yl)isophthalic (H_2L) acid as the organic linker, Ln-MOFs and mixed Ln-MOFs were produced by solvothermal method with the variation of molar ratios of Ln ions. P-XRD reveals that all the synthesized MOFs are isostructural, and from single-crystal XRD it is confirmed that EuL takes shape in monoclinic system with $C_{2/c}$ space group. When nine O atoms coordinate with five carboxylate O atoms from four diverse linkers, then a distorted tricapped triangular prismatic geometry is formed by the Eu^{3+} ion. In addition, two DMF molecules give two oxygen atoms and the nitrate ion gives the other two oxygen atoms. The bond distances of Eu–O remain in the range of 2.331–2.706 A°. The Eu^{3+} forms an SBU $\{Eu_2(COO)_4\}$ crossed over by four L ligands, and a 2D-layer framework is formed with the help of ligands and SBUs. Later on, those 2D frameworks were packed to form a 3D framework (Figure 2.16(b)) by the π–π interaction, and the crystal structure is displayed in Figure 2.16(a).

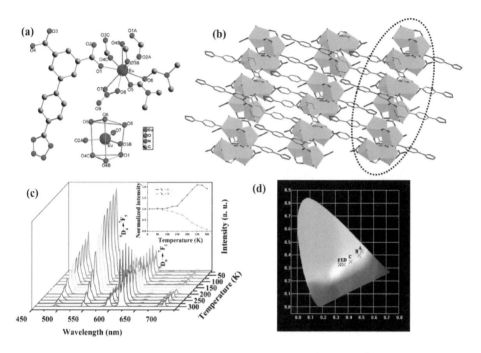

Figure 2.16: (a) Coordination environment of trivalent rare earth ion; inset shows the coordination polyhedron of Eu^{3+}. (b) Three-dimensional packing representation of EuL. (c) Temperature-dependent PL spectra of $Eu_{0.0878}Tb_{0.9122}L$ (10–300 K); the inset shows the normalized intensities of Tb^{3+} ($^5D_0 \rightarrow ^7F_2$) and Eu^{3+} ($^5D_4 \rightarrow ^7F_5$). (d) CIE chromaticity diagram for $Eu_{0.0748}Tb_{0.9252}L$ (A→F: λ_{exc} = 330–380 nm) [48]. Reproduced with permission from The Royal Society of Chemistry.

The temperature-dependent PL study was executed as far as intensity and lifetime in the temperature range from 10 to 300 K. EuL and TbL are not temperature sensitive because the luminescence intensities of Eu^{3+} and Tb^{3+} show a very few changes with increasing temperature. The $Eu_{0.0878}Tb_{0.9122}L$ exhibit completely different luminescence spectrum from EuL and TbL. In Figure 2.16(c), it is shown that there is decline in emission intensity of Tb^{3+} ion at 544 nm as well as lifetime of $^5D_4$ of Tb^{3+} reduces by 92%, while that of Eu^{3+} emission rises (no significant change in lifetime) with increase in temperature from 10–300 K. Thus, the observation supports the claim of energy move from Tb^{3+} to Eu^{3+} ion. This self-referencing temperature sensor uses the emission intensities of both Tb^{3+} (544 nm) and Eu^{3+} (614 nm) showing a good direct relationship ($T = 270.32 - 51.84\ I_{Tb}/I_{Eu}$) with temperature ranging from 75 to 250 K. The co-doped compound $Eu_{0.0748}Tb_{0.9252}L$ emits white light by the variation of the excitation wavelength from 330 to 380 nm. The gradual change in color from yellow to white is observed in the CIE chromaticity diagram as shown in Figure 2.16(d).

2.4.4.11 $Eu_{0.37}Tb_{0.63}$-BTC-a

In the year 2017, Wang, Zhao, Cui, Yang and Qian Huizhen reported a luminescent Ln-MOF which acted as a ratiometric thermometer to detect temperature in the high temperature range (313–473 K). To obtain $[Gd(C_9H_3O_6)H_2O]\cdot(H_2O)(DMF)$(Gd-BTC), $Gd(NO_3)_3\cdot 6(H_2O)$, 1,3,5-benzenetricarboxylate (H_3BTC) and hydrochloric acid were added in a solvent containing both DMF and water in it and then heated up to 80 °C by the solvothermal method. Similarly, with the variation of molar proportions of Eu and Tb ions, $Eu_{0.37}Tb_{0.63}$-BTC was formed. P-XRD confirms that the synthesized Ln-MOFs are isostructural, and Gd-BTC takes shape in the tetragonal group P_{4322}. The asymmetric unit is formed by a Gd atom, single BTC organic linker and a molecule of water, and the coordinated geometry is distorted pentagonal-bipyramidal. Two neighboring Gd^{3+} are connected by a carbonyl ligand having Gd–Gd distance of 4.71 A° and formed a condensed 3D framework (Figure 2.17(b)). Crystal structure of $Eu_{0.37}Tb_{0.63}$-BTC is highly thermostable (510 °C) and hence it can be used as a potential thermometer in the high temperature region. The emission and excitation spectra of EuBTC-a, TbBTC-a and $Eu_{0.37}Tb_{0.63}$-BTC were shown in Figure 2.17(c–e). EuBTC-a and TbBTC-a show their individual characteristic peaks, whereas $Eu_{0.37}Tb_{0.63}$-BTC-a shows both the representative peaks of Eu^{3+} at 612 nm and Tb^{3+} at 545 nm (λ_{exc} = 296 nm). As mentioned earlier, the emission intensity and lifetime are the two approaches to investigate the temperature-dependent luminescence properties. There is decline in the emission intensity of Tb^{3+}, while that of Eu^{3+} increments as temperature rises from 313 to 473 K (Figure 2.18(a)) (λ_{exc} = 296 nm). Here, the energy difference between the high triplet energized state of H_3BTC (26,504 cm^{-1}) and the accepting level of $Eu^{3+}(^5D_0$, 17,200 cm$^{-1})$ and $Tb^{3+}(^5D_4$, 20,500 cm$^{-1})$ is 9,304 cm^{-1} and 6,004 cm^{-1}, correspondingly. Such a high energy gap reflects the successful sensitization and energy transference from the ligand to the metal ions by phonon helped energy transference mechanism and also prevents the back-energy move.

As shown in Figure 2.18(b), $^5D_4$ of Tb^{3+} has a shorter lifetime, while $^5D_0$ of Eu^{3+} has a longer lifetime. From all the above discussion, it is found that the $Eu_{0.37}Tb_{0.63}$-BTC-a acts as a self-calibrating luminous temperature detector within the temperature range of 313–473 K with relative sensitivity of 0.68%/K at 313 K which decreases to 0.17%/K at 473 K.

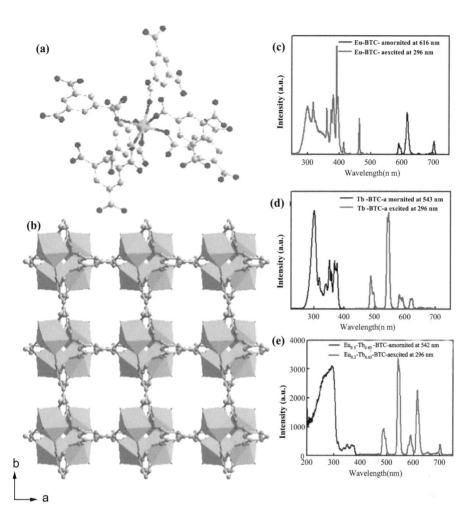

Figure 2.17: (a) Coordination environment of Gd^{3+}. (b) View of framework along the c-axis. Excitation and PL spectra of (c) Eu-BTC-a; (d) Tb-BTC-a; (e) $Eu_{0.37}Tb_{0.63}$-BTC-a. Reprinted from [44]. Copyright (2017), with permission from Elsevier.

2.4.4.12 TbMOF@3%Eu-tfac and TbMOF@7.3% Eu-tfac

Two new co-doped complexes having potential for use as thermometers (i.e., non-contact luminous temperature sensors) in the range of 200–400 K were created [22], by combining two unique proportions of a Eu^{3+} trifluoroacetylacetonate (tfac) with the smaller scale estimated $[Tb_2(bpydc)_3(H_2O)_3]\cdot nDMF$ particles by approved post-functionalization procedure. As TbMOFs are having excessively little pore to capture the tfac composite inside the MOF, that is why they are uniformly distributed over the

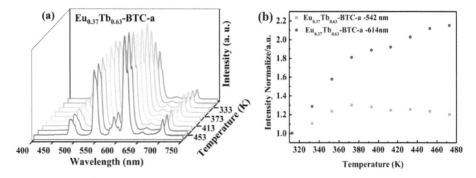

Figure 2.18: (a) PL spectra of $Eu_{0.37}Tb_{0.63}$-BTC-a in the temperature range of 313–473 K (λ_{exc} = 296 nm). (b) Temperature-dependent normalized intensities of $Eu_{0.37}Tb_{0.63}$-BTC-a ($^5D_4 \to {}^7F_5$ and $^5D_0 \to {}^7F_2$). Reprinted from [44]. Copyright (2017), with permission from Elsevier.

surface of the as-obtained TbMOF. The chosen TbMOF has a remarkable breathing conduct in structure. The 1D inorganic chains are formed by linking the TbO_x polyhedra (x = 7 or 8) through carboxylic sets. The chain was additionally connected by bpydc linkers to shape a 3D system. It is found that the structure of TbMOF stay unaltered after being post-functionalized with $Eu(tfac)_3 \cdot 2H_2O$.

The luminescence properties of the parental TbMOF material were examined at different temperatures (10–310 K). The most grounded emission intensity was seen at lower temperature (10 K), and as shown in Figure 2.19(a) there is gradual decrease in the emission intensity with the expansion in temperature range. The emission spectra for the TbMOF@3%Eu-tfac composite estimated over 200–400 K temperature run have been introduced in Figure 2.19(b), while the spectra of TbMOF@7.3%Eu-tfac are shown in Figure 2.19(c). In both cases, the Tb^{3+} emission intensity decreases much more than that of Eu^{3+} with the rise in temperature, signifying the procedure of energy transference from Tb^{3+} to Eu^{3+}, because of the thermal activation of nonradiative decay pathways. Thermometric parameters such as Δ, S_r and S_a were calculated to assess their usage as ratiometric detectors. The two MOFs could be considered as sensors in the wide temperature scope covering from 200 to 400 K. The former shows monotonic conduct in the range of 225–375 K, while the later shows that in the range of 200–325 K. The TbMOF@3%Eu-tfac indicated predominant conduct as a temperature detector based on intensity proportion. A greatest absolute sensitivity (S_a) of 0.069 K^{-1} (at 225 K) and relative sensitivity (S_r) of 2.59%/K (at 225 K) were determined for the former. The later demonstrated lower esteems of S_a of 0.012 K^{-1} (at 275 K) and S_r of 1.33%/K (at 325 K). As appeared by basically varying the measure of the Eu^{3+} complex doped with the TbMOF, one could acquire two different new ratiometric thermometer MOFs.

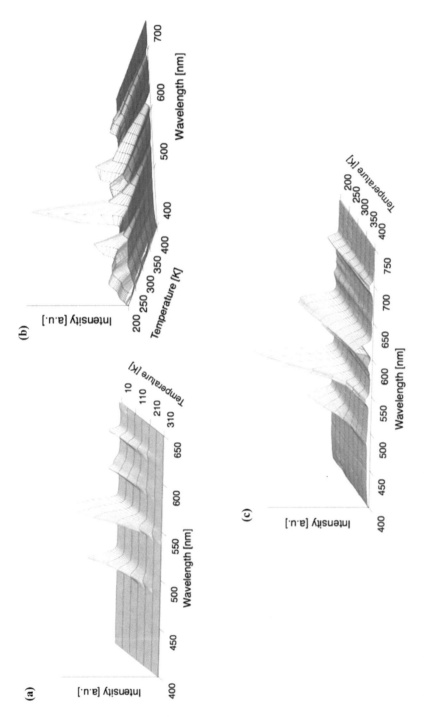

Figure 2.19: Temperature-dependent PL spectra of (a) TbMOF (10–310 K), (b) TbMOF@3%Eu-tfac and (c) TbMOF@7.3%Eu-tfac (200–400 K) [22]. Reproduced with permission from the Royal Society of Chemistry.

2.4.4.13 Eu$_{0.058}$Tb$_{0.942}$BPT

Zhang et al. [53] detailed a series of mixed Ln-MOFs by choosing biphenyl-tricarboxylate acid as an organic linker with reasonably high triplet excited level, among them Eu$_{0.058}$Tb$_{0.942}$BPT goes about as an incredible luminescent thermometer within the physiological temperature range. It acts as a thermometer over the whole physiological temperature range (20–80 °C), with the relative sensitivity greater than 1.5%/C. Solvothermal reaction of H$_3$BTP with Eu(NO$_3$)$_3$·6H$_2$O, Tb(NO$_3$)$_3$·6H$_2$O in a solution of DMF and H$_2$O brings about the establishment of colorless polyhedral crystals of mixed Ln-MOFs. These isostructural M′Ln-MOFs take shape in the monoclinic structure (space group P1) and four crystallographically free Tb^{3+} and BPT ligands form the asymmetric unit. Crystal structure of TbBPT is shown in Figure 2.20(a). The temperature-dependent emission spectra of Eu$_x$Tb$_{1-x}$BPT were executed (Figure 2.20(b)) for the potential utilization of Eu$_x$Tb$_{1-x}$BPT (x = 0.019, 0.058, 0.106) as a thermometer. The intensities of easily affected transitions, that is, $^5D_0 \rightarrow {}^7F_2$ for Eu^{3+} (615 nm) and $^5D_4 \rightarrow {}^7F_5$ for Tb^{3+} (545 nm) were taken as the ratiometric parameters. In the PL spectra, the emission intensity of the Tb^{3+} ions in Eu$_{0.058}$Tb$_{0.942}$BPT diminishes about 63%, which is higher than that reduced in the pure single rare earth-based MOF TbBPT (41%). Critically, the emission intensity of Eu^{3+} ions in Eu$_{0.058}$Tb$_{0.942}$BPT essentially increases to about 97%, while that of EuBPT diminishes to about 16% with the temperature increasing from 20 to 80 °C. The expanded thermal quenching of Tb^{3+} emission and the opposite inclination of Eu^{3+} emission might be ascribed to the energy transference from Tb^{3+} ions to it. Colorimetric temperature estimations are integral assets for body thermal mapping and intracellular temperature readout. CIE chromaticity diagram of Eu$_{0.058}$Tb$_{0.942}$BPT displayed in Figure 2.20(c) indicates the color change from green through orange and finally to red.

2.4.4.14 Eu$_{0.19}$Tb$_{0.81}$PDDI

Zhao et al. [57] reported a mixed Ln-MOF thermometer which is effective in the temperature range of 313–473 K as in the case of Eu$_{0.37}$Tb$_{0.63}$-BTC-a [44]. EuPDDI shows the favorable thermal stability of up to 380 °C. This type of MOF is effectively used for temperature detecting and diagnosis in microelectronics. This Tb/Eu mixed MOF system shows excellent temperature-sensitive PL behavior with a relative sensitivity of 0.19%/K at 313 K to 0.37%/K at 473 K, respectively. This value is higher than the other formerly announced Eu/Tb mixed MOFs. Eu(NO$_3$)$_3$·6H$_2$O, Tb(NO$_3$)$_3$·6H$_2$O, H$_4$PDDI, N,N'-dimethylformamide, toluene and HNO$_3$ solution were mixed in 20 mL screw-capped glass vial, after sealing that glass vial placed in an oven at 90 °C for 24 h. Colorless and transparent crystals were harvested after cooling the mixture to room temperature naturally, wash away with DMF and ethanol commonly and afterward dried out in air. From powder crystal, XRD affirmed that all the three Ln-MOFs, that

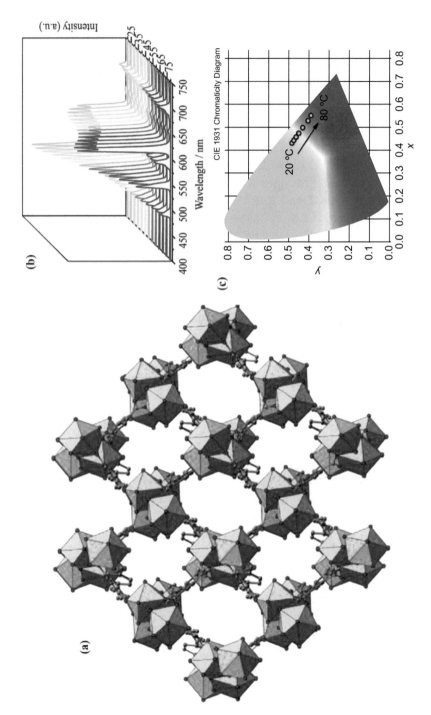

Figure 2.20: (a) Crystal structure of TbBPT. (b) PL spectra of $Eu_{0.058}Tb_{0.942}BPT$ in the temperature range of 20–80 °C (λ_{exc} = 325 nm). (c) CIE chromaticity diagram of $Eu_{0.058}Tb_{0.942}BPT$. Reprinted from [53]. Copyright (2018), with permission from Elsevier.

is, EuPDDI, TbPDDI and $Eu_{0.19}Tb_{0.81}PDDI$ are isostructural, and EuPDDI crystallizes in the monoclinic structure (space group $C_{2/c}$).

The free H_4PDDI ligand displays a broad blue emission band around 477 nm (λ_{exc} = 396 nm) which is expected to be the intra-ligand π–π* electronic progression. In case of EuPDDI and TbPDDI, both show the typical characteristic peaks of Eu^{3+} ions at 591, 615, 653 and 700 nm due to $^5D_0 \rightarrow \,^7F_{1-4}$ transitions and the characteristic peaks of Tb^{3+} at 489, 545, 584 and 620 nm due to $^5D_4 \rightarrow \,^7F_{6-3}$ transitions, respectively. However, the mixed Ln-MOF $Eu_{0.19}Tb_{0.81}PDDI$ simultaneously shows the characteristic peaks of both the metal ions, which implies that the organic ligand H_4PDDI go about as a superb antenna chromophore to successfully sensitize the luminescence of both Tb^{3+} and Eu^{3+} ions (Figure 2.21(b)). In addition, the luminescence of Eu^{3+} ion is further sensitized by the energy move from Tb^{3+} to Eu^{3+}, which is mainly dependent on temperature, that is, as temperature increases from 313 to 473 K, the intensity of Eu^{3+} increases bit by bit while that of Tb^{3+} declines (Figure 2.21(c)). Thus, the mixed Ln-MOF $Eu_{0.19}Tb_{0.81}PDDI$ can be utilized as the ratiometric parameter for detecting temperature by taking the proportion between the luminescence intensity of Tb^{3+} and Eu^{3+} ions. With the temperature range from 313 to 473 K, the emission color of $Eu_{0.19}Tb_{0.81}PDDI$ progressively moves from white to red.

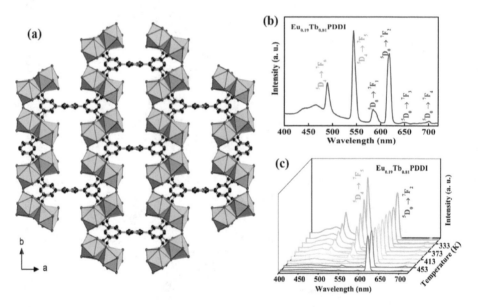

Figure 2.21: (a) View of framework along the c-axis (sky blue polyhedron, Eu; red, O; black, C; blue, N). (b) Excitation and emission spectra of $Eu_{0.19}Tb_{0.81}PDDI$. (c) Temperature-dependent PL spectra of $Eu_{0.19}Tb_{0.81}PDDI$ (T = 313–473 K) [57]. Reproduced with permission from the Royal Society of Chemistry.

2.4.4.15 Tb$_{0.95}$Eu$_{0.05}$FTPT

In the year 2018, Zhao, Yue, Zhang, Jiang and Qian reported a fluorine-altered tetracarboxylic acid ligand, 2′-fluoro-[1,1′:4′,1″-terphenyl]-3,3″,5,5″-tetracarboxylic acid (H$_4$FTPTC), which was structured and used to build an Ln-MOF Tb$_{1-x}$Eu$_x$FTPTC (x = 0.05, 0.1, 0.2) for cryogenic temperature sensing (25–125 K). The energy-level distribution of the organic linker can be affected by the alteration of [1,1′:4′,1″-terphenyl]-3,3″,5,5″-tetracarboxylic acid with fluorine atom; as a result, T$_1$ of the ligand becomes near to the accepting level of the metal ions which satisfy temperature detection in the cryogenic regime. With the help of Suzuki coupling reaction, followed by hydrolysis and then acidification, the organic linker H$_4$FTPTC was synthesized (as shown in Scheme 2.2).

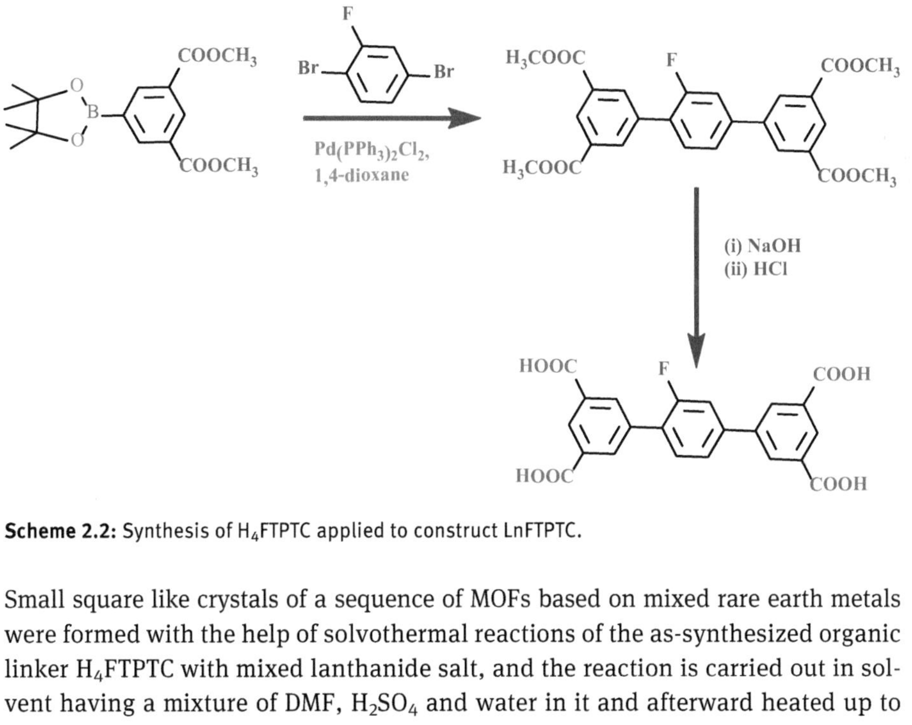

Scheme 2.2: Synthesis of H$_4$FTPTC applied to construct LnFTPTC.

Small square like crystals of a sequence of MOFs based on mixed rare earth metals were formed with the help of solvothermal reactions of the as-synthesized organic linker H$_4$FTPTC with mixed lanthanide salt, and the reaction is carried out in solvent having a mixture of DMF, H$_2$SO$_4$ and water in it and afterward heated up to 90 °C for 3 days. All the prepared Ln-MOFs are isostructural and it was found that TbFTPTC crystallizes in the monoclinic structure (space group C$_{2/m}$) and forms a 3D framework built of binuclear subunits of [Tb$_2$(μ$_2$-COO)$_4$(COO)$_2$]. The binuclear subunits are formed by the crossing over of two crystallographically autonomous Tb atoms (spatially separated at a distance of 4.458 A°) by carbonyl units of FTPTC, as shown in Figure 2.22(a). The coordination geometries around both the Tb centers are distorted tricapped trigonal prism and the bond distances of Tb–O bonds in TbFTPTC extend from 2.264 to 2.564 A°. The MOFs are thermally stable up to 310 °C, and this was confirmed from thermogravimetric analysis (TGA).

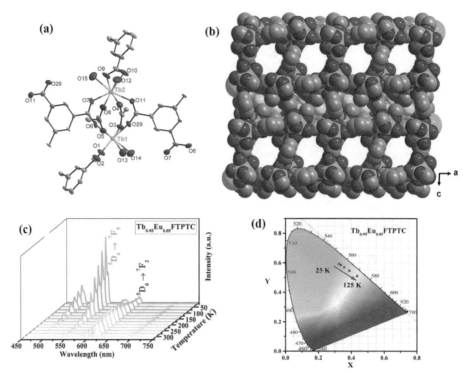

Figure 2.22: (a) Coordination environment of TbFTPTC. (b) Three-dimensional crystal packing along the b-axis (Tb^{3+}, green; C, gray; F, cyan; O, red). (c) PL spectra of Tb$_{0.95}$Eu$_{0.05}$FTPTC (T = 25–300 K). (d) CIE chromaticity diagram of Tb$_{0.95}$Eu$_{0.05}$FTPTC showing color variation from green (x = 0.3255, y = 0.5912) at 25 K to (x = 0.4351, y = 0.5095) at 125 K. Reprinted with permission from [59]. Copyright (2018) American Chemical Society.

The emissions displayed by TbFTPTC for Tb^{3+} at 490, 543, 588 and 621 nm are due to $^5D_4 \rightarrow {}^7F_{6-3}$ transitions and that displayed by EuFTPTC at 580, 592, 613, 653 and 703 nm are due to $^5D_0 \rightarrow {}^7F_{0-4}$ transitions of Eu^{3+} ion. Whereas the characteristic emissions of both Tb^{3+} and Eu^{3+} are concurrently shown by the mixed Ln-MOFs Tb$_{1-x}$Eu$_x$FTPTC. But the emission band of H$_4$FTPTC is not detected, which plainly infers that it goes about as an appropriate chromophore for sensitizing both the Ln^{3+} ions. The temperature-dependent PL spectra of Tb$_{0.95}$Eu$_{0.05}$FTPTC (Figure 2.22(c)) showed that the intensity of Tb^{3+} decreases as temperature increases, and the reverse is true for Eu^{3+}. Because of energy transference from Tb^{3+} to Eu^{3+}, the lifetime of Tb^{3+} dramatically decreases in Tb/Eu-MOF than that of pure MOF, while there is no significant decrease in the lifetime of Eu^{3+}. The relative sensitivity of Tb$_{0.95}$Eu$_{0.05}$FTPTC is found to be 0.91%/K at 25 K, then increases to 9.11%/K at 125 K. Similarly, Tb$_{0.90}$Eu$_{0.10}$FTPTC and Tb$_{0.80}$Eu$_{0.20}$FTPTC show the relative sensitivity scope from 1.29 to 36.90%/K and from 0.93 to 12.94%/K, which implies that these two MOFs also act as luminescent thermometers. CIE chromaticity diagram in Figure 2.22(d) shows the color coordinate change from green to yellow-green.

2.4.4.16 Tb$_{0.93}$Eu$_{0.07}$-BODSDC

A benzophenone-3,3′-disulfonyl-4,4′-dicarboxylic acid (H$_4$-BODSDC) is used as an organic ligand to prepare an Ln-MOF Tb$_{0.93}$Eu$_{0.07}$-BODSDC, which acts as a temperature sensor covering the range from 100 to 300 K showing a highest sensitivity of 0.23%/K at 300 K [54]. Scheme 2.3 shows a two-step reaction sequence for the preparation of the organic ligand H$_4$-BODSDC, and then {(H$_3$O)[Ln(BODSDC)-(H$_2$O)$_2$]}$_n$ (Ln = Tb, Eu and Gd) was prepared by the solvothermal method. Tb$_{0.93}$Eu$_{0.07}$-BODSDC was prepared by taking nitrates of rare earths [Tb(NO$_3$)$_3$/Eu(NO$_3$)$_3$] with mole proportion of 0.95/0.05, and H$_4$-BODSDC in CH$_3$CN/H$_2$O.

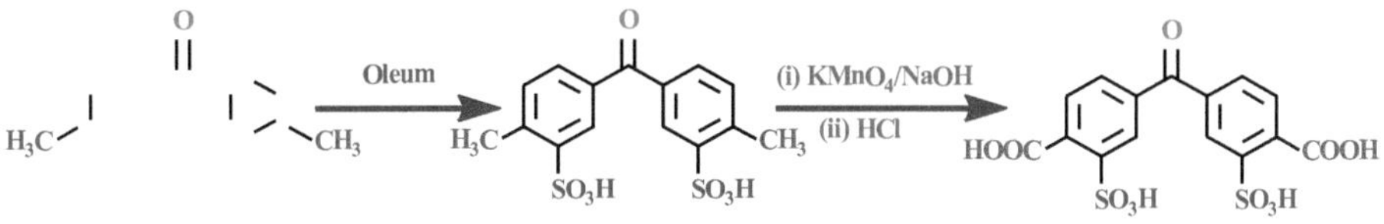

Scheme 2.3: Synthesis of H$_4$-BODSDC.

From single-crystal XRD, it is confirmed that Ln-BODSDC takes shape in a monoclinic structure (space group P$_{2/c}$) and has a 3D system comprising 1D channels. The 1D chains are formed when the lanthanide centers are crossed over by the carboxylate sets of BODSDC^{4-} ligands, which are then associated by the BODSDC^{4-} linkers to result in a 3D-framework (Figure 2.23(a)). The bond lengths of Ln–O vary from 2.321(2) to 2.4561(19) A°, and the Ln–Ln separation is 4.6535(1) A°. From the temperature-dependent PL study, it is confirmed that the Tb^{3+} and Eu^{3+} ions in Tb$_{0.93}$Eu$_{0.07}$-BODSDC are effectively sensitized by the BODSDC^{4-} organic linker with a proper triplet energized state. It was observed that the characteristic peaks of Tb^{3+} at 489, 544, 584, 622 and 652 nm are due to $^5D_4 \rightarrow {}^7F_{(J = 6, 5, 4, 3, 2)}$ transitions and that of Eu^{3+} at 593, 616, 653, 697 and 736 nm are due to $^5D_0 \rightarrow {}^7F_J$ $_{(J = 1, 2, 3, 4, 5)}$ transitions (λ_{exc} = 368 nm). As the temperature increases from 100 to 300 K, there is an increase in the luminescence intensity of Eu^{3+}, while that of Tb^{3+} declines (Figure 2.23(b)), because of the transference of energy between the rare earth ions within the framework.

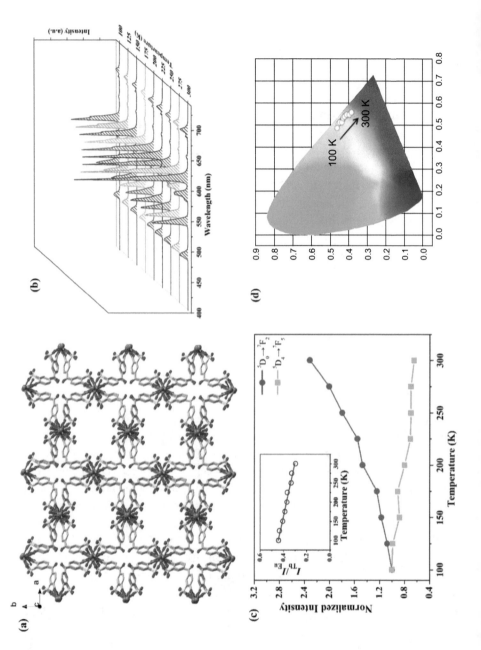

Figure 2.23: (a) Three-dimensional framework of Tb(1). (b) Temperature-dependent PL spectra of Tb$_{0.93}$Eu$_{0.07}$-BODSDC under excitation at 368 nm. (c) Temperature-dependent normalized intensities of $^5D_4 \rightarrow {}^7F_5$ (536–556 nm) and $^5D_0 \rightarrow {}^7F_2$ (604–633 nm) transitions; the inset shows the intensity proportion. (d) CIE chromaticity diagram for Tb$_{0.93}$Eu$_{0.07}$-BODSDC. Reprinted with permission from [54]. Copyright (2018) American Chemical Society.

2.4.4.17 Eu$_{0.015}$Tb$_{0.985}$-tatab

Solvothermal reaction of 4,4′,4″-s-triazine-1,3,5-triyltri-p-aminobenzoic acid (H$_3$tatab, a semirigid tripodal carboxylate ligand) with nitrates of rare earths [Tb(NO$_3$)$_3$/Eu(NO$_3$)$_3$] in mixed diluents of N-methyl pyrrolidone (NMP) and water yields novel 3D breathing Ln-MOFs. The MOF includes pure rare earth Eu-tatab, Tb-tatab and bimetallic Eu$_{0.015}$Tb$_{0.985}$-tatab (having the breathing capacity to precipitously discharge guests and keep up a similar topology with the reduction of cell volume) with 1D channels [51]. Eu$_{0.015}$Tb$_{0.985}$-tatab shows higher sensitivity as thermometer depending on the intensity proportion between the emissions of both Tb^{3+} and Eu^{3+} at 545 and 614 nm, respectively. This ratiometric thermodetector can be utilized in an extensive temperature range from 90 to 300 K. Specifically, not quite the same as reported in Eu/Tb luminescent thermometers, for example, DMBDC [11], the luminescence color of Eu$_{0.015}$Tb$_{0.985}$-tatab steadily blueshift, as opposed to redshift, with increasing temperature. All the synthesized MOFs were isomorphous, which is known from single-crystal XRD and IR spectra. Taking Eu-tatab as a representative, it is found that it takes shape in the monoclinic structure (space group P21/c), where two Eu^{3+} ions and a single tatab organic linker which is triply deprotonated are coordinated to form the asymmetric unit. The crystal structure is shown in Figure 2.24(a, b).

Eu-tatab′ displayed the emission of tatab and trademark red emissions of Eu^{3+}, while Tb-tatab′ only showed trademark green emission of Tb^{3+}, signifying the facile energy transference from tatab to Tb^{3+}. Eu$_{0.015}$Tb$_{0.985}$-tatab′ shows the representative emission of both Tb^{3+} and Eu^{3+}. The temperature-dependent PL spectra of the abovementioned MOFs are displayed in Figure 2.24(c–e). The luminescent intensity of Eu-tatab′ diminished bit by bit with an expansion in temperature from 90 to 300 K, while that of Tb-tatab′ does not show an undeniable change. When there is an increase in temperature from 90 to 300 K, then in case of doped mixed Eu$_{0.015}$Tb$_{0.985}$-tatab′ all characteristic transitions $^5D_4 \rightarrow {}^7F_{6-4}$ (Tb^{3+}) and $^5D_0 \rightarrow {}^7F_{1-4}$ (Eu^{3+}) were lessened/diminished with the degree of decline of Eu^{3+} being quicker than that of Tb^{3+}, particularly in the range of 230–300 K. Similarly, their lifetime also decreases significantly with the increase in temperature.

2.4.4.18 Eu$_{0.0316}$Tb$_{0.9684}$BTPTA

By taking 4,4′,4″-(benzene-1,3,5-triyltris(1H-pyrazole-3,1-diyl))tribenzoic acid (H$_3$BTPTA) as an organic linker, an Eu/Tb-mixed rare-earth-based CP Eu$_{0.0316}$Tb$_{0.9684}$BTPTA [64] was reported. This Tb/Eu mixed CP acts as a ratiometric luminescent thermometer having relative sensitivity of 0.45–5.12%/K in the temperature range of 25–225 K. A sequence of rare earth metal-based CP LnBTPTA (Ln = Eu, Tb, Eu$_x$Tb$_{1-x}$) was produced by the solvothermal process. P-XRD confirmed that all the synthesized Ln-CPs are isostructural. By taking TbBTPTA, for instance, single-crystal

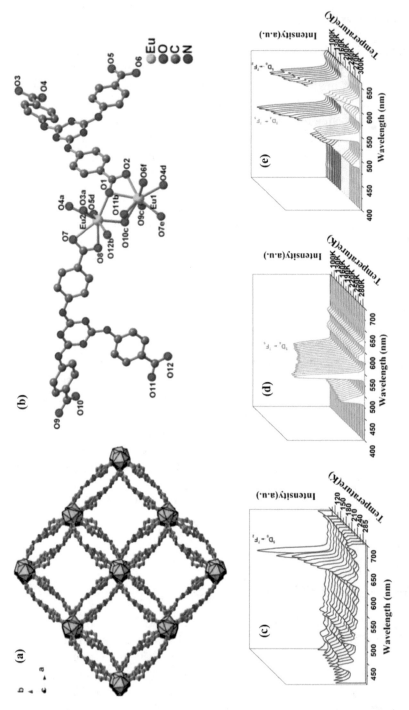

Figure 2.24: (a) Three-dimensional framework along the c-axis. (b) Coordination environment of Eu^{3+}. PL spectra of (c) Eu-tatab′, (d) Tb-tatab′ and (e) $Eu_{0.015}Tb_{0.985}$-tatab′. Reprinted with permission from [51]. https://pubs.acs.org/doi/full/10.1021/acsomega.8b00199.

XRD affirmed that it takes shape in the monoclinic structure (space group I2/a) and the asymmetric unit cell is composed of seven crystallographically autonomous Tb^{3+} ions and seven crystallographically free $BTPTA^{3-}$ organic linkers (Figure 2.25(a)), which is very uncommon. In TbBTPTA, the bond distances of Tb–O stay in the scope of 2.278 (6)–2.582(3) A°. As shown in Figure 2.25(b), Tb^{3+} ions composed of trigonal organic linkers forming 2D thick layers having voids in a twofold Z-shaped layered framework were created with two types of windows: one is a rhombic form window along the b-axis of 16×3 A^{o2} and the other one is a cucurbit-like window with a diameter of about 4 A°. $Eu_{0.0316}Tb_{0.9684}BTPTA$ concurrently displayed the characteristic emission peaks of both the metal ions (λ_{exc} = 377 nm); out of them $^5D_0 \rightarrow {}^7F_2$ (Eu^{3+}, 614 nm) and $^5D_4 \rightarrow {}^7F_5$ (Tb^{3+}, 545 nm) are the most prominent of all transitions. H_3BTPTA having a triplet state (T_1) energy level of 27,473 cm^{-1} acts as an antenna chromophore to efficiently sensitize both Tb^{3+} and Eu^{3+} ions, with the absence of ligand peak at room temperature. Also, the emission of Eu^{3+} ions in $Eu_{0.0316}Tb_{0.9684}BTPTA$ can be further sensitized by the energy transference from Tb^{3+} ions. Within the temperature extent of 25–275 K, the PL properties of the Ln-MOF ($Eu_{0.0316}Tb_{0.9684}BTPTA$) were investigated to ensure its capability to be utilized as a ratiometric temperature detector. When temperature increases from 25 to 275 K, because of the thermal activation of non-radiative decay pathways, there is a decrease in the emission intensity of Eu^{3+} and Tb^{3+} in EuBTPTA and TbBTPTA, respectively. However, bimetallic mixed CP $Eu_{0.0316}Tb_{0.9684}BTPTA$ shows a completely extraordinary temperature-dependent luminescence performance than those of single metal-based CPs EuBTPTA and TbBTPTA. In the cryogenic region, $Eu_{0.0316}Tb_{0.9684}BTPTA$ shows a stronger emission intensity for Tb^{3+} than that of Eu^{3+}.

Because of the energy transference process from Tb^{3+} to Eu^{3+} (Figure 2.25(c)), a rapid decline in the emission intensity of Tb^{3+} (543 nm) occurs and that of Eu^{3+} expands first at that point, and drops later with the increment in temperature. Similarly, the lifetime of the CP, $Eu_{0.0316}Tb_{0.9684}BTPTA$ decreases with increasing temperature. The effectiveness of energy transference from Tb^{3+} to Eu^{3+} can be determined from the lifetime of the benefactor luminescence (using the relation $E = 1 - \tau/\tau_0$, where τ and τ_0 are the lifetimes of the contributor (i.e., Tb^{3+}) in the existence and nonattendance of the acceptor (Eu^{3+}). The maximum sensitivity of $Eu_{0.0316}Tb_{0.9684}BTPTA$ is 5.12%/K (Figure 2.25(d)).

2.4.4.19 $Tb_{0.95}Eu_{0.05}$cpna and $Tb_{0.95}Eu_{0.05}$bpydc

Two isostructural trivalent rare-earth-based MOFs named as $Tb_{0.95}Eu_{0.05}$cpna and $Tb_{0.95}Eu_{0.05}$bpydc (Ln-MOFs) were reported, and their sensitivities were compared [58]. For this, two dicarboxylate ligands containing pyridine, that is, {6-(4-carboxyphenyl)-nicotinic acid and [2,2'-bipyridine]-5,5'-dicarboxylic acid}, which are structurally similar (Scheme 2.4) were taken into consideration. These two organic linkers containing different number of pyridine units offer various minimal triplet state

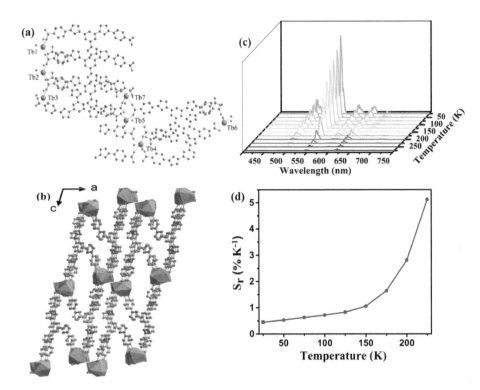

Figure 2.25: (a) Asymmetric unit of TbBTPTA. (b) View of crystal structure along the b-axis. (c) Temperature-dependent PL spectra of $Eu_{0.0316}Tb_{0.9684}$BTPTA (T = 25–275 K) under excitation at 377 nm. (d) Relative sensitivity of $Eu_{0.0316}Tb_{0.9684}$BTPTA. Reprinted from [64]. Copyright (2018), with permission rom Elsevier.

energies, which influence the sensitization of organic linkers to Eu^{3+} and Tb^{3+} ions. They show an S-formed response for sensing temperature in the scope of 25–300 K. Solvothermal reaction of the organic linkers H_2cpna/H_2bpydc with the nitrates of the rare earths $Tb(NO_3)_3$/$Eu(NO_3)_3$ having mole proportion of 0.95/0.05 was added to the mixed solvent having mixtures of Dimethylacetamide (DMA), chlorobenzene, and methylbenzene in it and then heated for 3 days at 115 °C to form the block-like crystals of respective mixed Ln-MOFs $Tb_{0.95}Eu_{0.05}$cpna and $Tb_{0.95}Eu_{0.05}$bpydc. P-XRD reveals that all the synthesized MOFs are isostructural. By considering the case of Eucpna, single-crystal XRD confirmed that it takes shape in the monoclinic structure having P21/n space group. A slightly distorted bicapped trigonal prismatic geometry is formed around the Eu^{3+} ion (Figure 2.26).

The bond distances of Eu–O (carboxylic O) range from 2.320(6) to 2.351(6) Å and the spatial distance between neighboring Eu^{3+} ions is 4.7220(7) Å. The cpna organic ligands develop a 3D system by connecting infinite Eu-carboxylate chains in an equal way with one another.

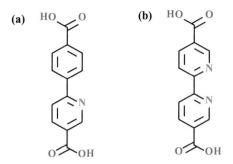

Scheme 2.4: Structure of H$_2$cpna and H$_2$bpydc.

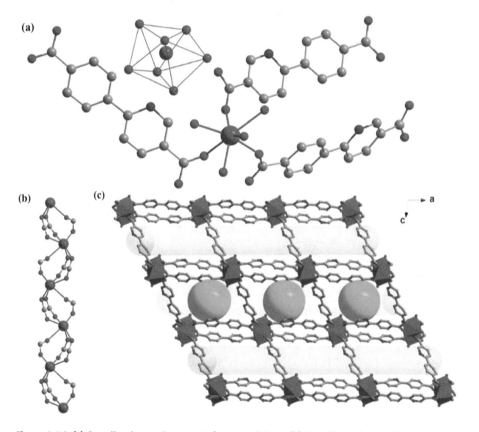

Figure 2.26: (a) Coordination environment of rare earth ions. (b) One-dimensional helical chain along the *b*-axis made by infinite Eu^{3+}-carboxylate units. (c) Two groups of interpenetrated and perpendicular 1D rhombic channels along the *a*- and *b*-axes (blue, Eu; gray, C; red, O; blue, N). Reprinted with permission from [58]. Copyright (2019) American Chemical Society.

Tbcpna (or Eucpna) and Tbbpydc (or Eubpydc) show their corresponding characteristic emissions when excited at 328 and 339 nm, individually. While the emission peaks of the organic ligands, H_2cpna and H_2bpydc, entirely vanish in their respective MOFs, representing their excellent sensitization property for Tb^{3+} and Eu^{3+} ions. When excited at room temperature under UV light, both the doped M'Ln-MOFs, $Tb_{0.95}Eu_{0.05}$cpna and $Tb_{0.95}Eu_{0.05}$bpydc, concurrently show the distinctive luminescence of Tb^{3+} and Eu^{3+} ions. The temperature-dependent PL spectra of $Tb_{0.95}Eu_{0.05}$cpna and $Tb_{0.95}Eu_{0.05}$bpydc are shown in Figure 2.27(a, b). In the PL spectrum of $Tb_{0.95}Eu_{0.05}$cpna, when there is expansion in temperature from 25 to 150 K, then it is found that there is a decrease in the intensity peak of Tb^{3+} by 0.45%/K. At the same time, there is huge improvement in the emission intensity of Eu^{3+}, because of energy transference from Tb^{3+} to Eu^{3+} (which depends on the thermally determined phonon-aided energy transference mechanism). $Tb_{0.95}Eu_{0.05}$bpydc also shows the similar behavior, but with different extent of variation in the corresponding emissions. It is seen that the sensitivity of $Tb_{0.95}Eu_{0.05}$bpydc is more as compared to $Tb_{0.95}Eu_{0.05}$cpna, due to the presence of two pyridine units in H_2bpydc. The sensitivity of $Tb_{0.95}Eu_{0.05}$cpna is found to be 2.55% at 131 K, while that of $Tb_{0.95}Eu_{0.05}$bpydc is 2.59% at 179 K (Figure 2.27(c)).

2.4.4.20 $Eu_xTb_{1-x}L$

Three dual-emitting Ln-MOF hybrids $Tb_{1-x}Eu_xL$ (x = 0.05, 0.1 and 0.2) were created as ratiometric MOF temperature sensors, considering the energy transference mechanism from $Tb^{3+} \rightarrow Eu^{3+}$, which makes the thermometers increasingly delicate and exact for temperature sensing [50]. From these three Ln-MOF hybrids, $Tb_{0.9}Eu_{0.1}L$ can be used as a thermometer for sensing temperature in a region from the physiological (303–320 K) to high-temperature range (320–423 K) having a maximum sensitivity (S_m) of 1.75%/K at 423 K. This S_m value is about 5 times as compared to the first co-doped Ln-MOF temperature sensor $Eu_{0.0069}Tb_{0.9931}$-DMBDC reported by Cui et al. [11], and 13 times of the first co-doped Ln-MOF nanothermometer reported by Cadiau et al. [8]. $Tb_{1-x}Eu_xL$ hybrid Ln-MOFs were prepared by the solvothermal reaction of ligand H_2L and mixtures of $Eu(NO_3)_3 \cdot 6H_2O$ and $Tb(NO_3)_3 \cdot 6H_2O$ adding in a mixed solvent of DMF/EtOH/H_2O, which is then packed in a glass bottle. After that, the glass bottle containing reaction mixture was heated for 3 days at 85 °C. The subsequent crystals were dried normally after gathered by filtration and then washed 3 times with DMF. Figure 2.28(a) shows the 2D framework structure of $Tb_{1-x}Eu_xL$. It takes shape in the triclinic structure with space group P1 and the asymmetric unit is formed by the combination of Eu^{3+}/Tb^{3+} ion, two organic linkers (L) and an anion of NO_3^-, which is confirmed by single-crystal XRD. Every Eu^{3+}/Tb^{3+} can form a nine-coordinated center, which forms a distorted tricapped triangular prismatic geometry by combining with 5L ligands. TGA analysis confirmed the higher thermostability of the synthesized MOFs.

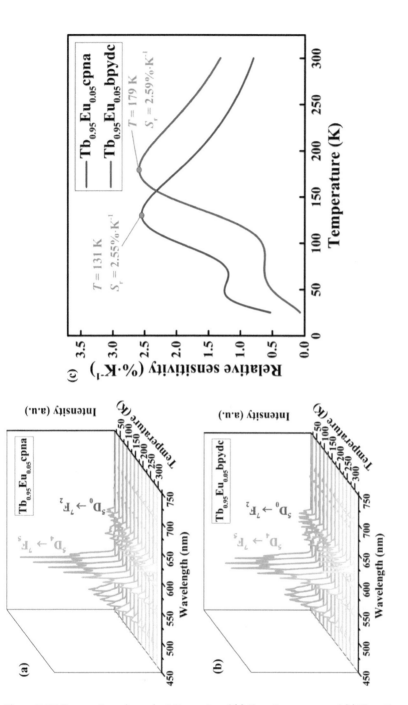

Figure 2.27: Temperature-dependent PL spectra of (a) $Tb_{0.95}Eu_{0.05}$cpna and (b) $Tb_{0.95}Eu_{0.05}$bpydc (T = 25–300 K). (c) Relative sensitivity of both the Ln-MOFs (T = 25–300 K). Reprinted with permission from [58]. Copyright (2019) American Chemical Society.

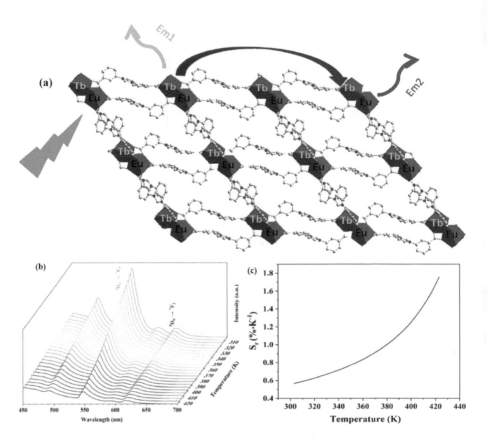

Figure 2.28: (a) Two-dimensional framework of $Tb_{1-x}Eu_xL$. (b) PL spectra of $Tb_{0.9}Eu_{0.1}L$ (303–423 K). (c) The relative sensitivity values (S_r) of $Tb_{0.9}Eu_{0.1}L$. Reprinted from [50]. Copyright (2020), with permission from Elsevier.

In case of $Tb_{0.95}Eu_{0.05}L$, the sharp characteristic luminescent intensity is inapparent at 617 nm for Eu^{3+} ($^5D_0 \rightarrow {}^7F_2$ transition), while in case of $Tb_{0.8}Eu_{0.2}L$ that becomes too low for Tb^{3+} at 547 nm ($^5D_4 \rightarrow {}^7F_5$ transition); so, $Tb_{0.9}Eu_{0.1}L$ was selected as the best representative. The temperature-dependent PL spectra of $Tb_{0.9}Eu_{0.1}L$ are shown in Figure 2.28(b). When temperature increases from 303 to 423 K, there is a decrease in the luminescent intensity of Tb^{3+} (547 nm, $^5D_4 \rightarrow {}^7F_5$) in the co-doped system, while that of Eu^{3+} (617 nm, $^5D_0 \rightarrow {}^7F_2$) almost have no significant decrease. This phenomenon is because of the thermally driven energy transference from $Tb^{3+} \rightarrow Eu^{3+}$. The intensity ratio of both the ions, that is, Tb^{3+} (547 nm) and Eu^{3+} (617 nm) is commonly used as the parameter for measuring temperature in the thermometers.

2.4.4.21 [(CH$_3$)$_2$NH$_2$]Eu$_{0.036}$Tb$_{0.964}$BPTC

A stable Eu/Tb-bimetallic MOF ([(CH$_3$)$_2$NH$_2$]Eu$_{0.036}$Tb$_{0.964}$BPTC), which acts as an excellent luminescent thermometer for detecting temperature in an extensive range of temperature from 77 to 377 K, based on the distinguished characteristic emissions of lanthanides [34] was reported. This temperature detector shows good straight response with temperature within the range of 220–310 K, having a maximum relative sensitivity (S_m) of 9.42%/K at 310 K. By the solvothermal method, three types of Ln-MOFs were produced, among them two are having single lanthanide metal ion ([(CH$_3$)$_2$NH$_2$]EuBPTC, [(CH$_3$)$_2$NH$_2$]TbBPTC) and the other one is mixed Ln-MOF ([(CH$_3$)$_2$NH$_2$]Eu$_{0.036}$Tb$_{0.964}$BPTC). These three compounds are isostructural, confirmed by single-crystal XRD analyses. Considering [(CH$_3$)$_2$NH$_2$]EuBPTC, it solidifies in the trigonal structure (space group P3$_1$), where a BPTC ligand and a crystallographically free Eu^{3+} ion are coordinated to form the asymmetric unit. The decarbonylation reaction of DMF molecules generates the [Me$_2$NH$_2$]$^+$ counterions, and the framework is balanced by two [Me$_2$NH$_2$]$^+$ units. The coordination environment and 3D framework of [(CH$_3$)$_2$NH$_2$]EuBPTC is depicted in Figure 2.29(a, b), respectively. The temperature-dependent PL properties of [(CH$_3$)$_2$NH$_2$]EuBPTC and [(CH$_3$)$_2$NH$_2$]TbBPTC enabled the development of mixed Ln-MOF [(CH$_3$)$_2$NH$_2$]Eu$_{0.036}$Tb$_{0.964}$BPTC. The ligand having a triplet state (T$_1$) energy level is located at 29,070 cm^{-1}, which can make the ligand to effectively sensitize both the trivalent rare earth ions. In Figure 2.29(c), the PL spectra show that the temperature-dependent PL properties of Tb^{3+} at 545 nm and Eu^{3+} at 614 nm are completely different from that of respective single lanthanide compounds. The emission intensity of both the Ln^{3+} ions stays unaltered with the rise in temperature in the lower temperature region (77–167 K), while that in the range of 167–377 K is reduced by about 89% for Tb^{3+} ions and increases by 400% for Eu^{3+} ions (Figure 2.29(d)) regardless of the small content of Eu^{3+}. This dramatic change is because of the thermally determined energy transference from Tb^{3+} to Eu^{3+}, which can act as an outstanding ratiometric self-referencing luminescent thermometer.

Similarly, in the mixed Ln-MOFs, the lifetime of Tb^{3+} ions became shorter, whereas it became longer in the case of Eu^{3+} ions (Figure 2.29(e)). The temperature-dependent luminescence colors also changed from green, through yellow and finally to red with expanding temperature, which can be seen by the naked eye or camera (Figure 2.29(f)).

2.4.4.22 Eu$_{0.05}$Tb$_{1.95}$-PDC

Xiao Zhou et al. conveyed a progression of rare earth metal-based organic frameworks, in particular, [Ln$_2$PDC$_3$-(H$_2$O)]H$_2$O (Ln = Eu, Tb, Eu$_x$Tb$_{2-x}$) with high-efficiency luminescence and long fluorescence lifetime have been prepared through

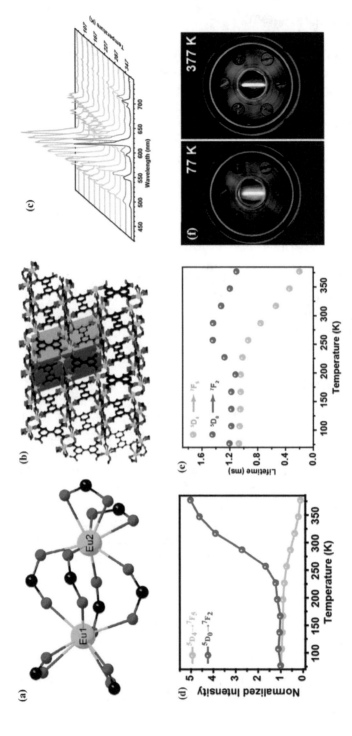

Figure 2.29: (a) Coordination environment of Eu^{3+} unit which is binuclear. (b) Three-dimensional structures along the a-axis. (c) Solid-state PL spectra of $[((CH_3)_2NH_2]Eu_{0.036}Tb_{0.964}BPTC$ (77–377 K). Temperature-dependent (d) normalized intensity and (e) lifetime of $^5D_4 \rightarrow {}^7F_5$ and $^5D_0 \rightarrow {}^7F_2$ transitions (77–377 K) for $Eu_{0.036}Tb_{0.964}BPTC$. (f) Optical photographs of the luminous sample [34]. Reproduced with permission from the Royal Society of Chemistry.

the solvothermal method by taking pyridine-3,5-dicarboxylic (H_2PDC) acid as the organic ligand [62]. By altering the range of Eu/Tb in the as-prepared Eu_xTb_{2-x}-PDC complex tunable emission shading from green, through yellow to red is figured out. Among them, $Eu_{0.05}Tb_{1.95}$-PDC complex showed a potential solid-state material for a luminous temperature detector in the scope of 293–333 K and having the temperature resolution under 0.16 K. Taking Eu-PDC as an example, these complexes take shape in the monoclinic structure (space group P21/n). The asymmetric unit of Eu-PDC encompasses a homodinuclear structure; to be specific, it comprises two rare earth centers of europium atoms (Eu1, Eu2), which is not quite the same as the reported written works (Figure 2.30). Eu1 is eight-coordinated having a distorted trigonal dodecahedron geometry, while Eu2 is nine-coordinated having the coordination polyhedron as a distorted tricapped trigonal prism geometry around Eu2.

When excited at 284 nm, both Tb^{3+} and Eu^{3+} show sharp peaks are credited to their characteristic transitions. This outcome confirms that there is effective energy transference happens between the organic linker and the rare earth ions, and PDC^{2-} sensitize both the lanthanide ions (Tb^{3+}, Eu^{3+}), which acts as a fantastic antenna chromophore for them.

Simultaneously, through checking at 611 nm (Eu^{3+}: $^5D_0 \rightarrow {}^7F_2$), the emission peaks of Tb^{3+} are seen in the excited range of Eu-Tb-PDC, as shown in Figure 2.31(a). Obviously, it confirms different possibilities of energy transference to Eu^{3+} from Tb^{3+}. By checking the $^5D_4$ energized state of Tb^{3+} (at 542 nm) (Figure 2.31(b)), the lifetime decay curves are verified. The fluorescence lifetime differently decreases with the expansion of Eu^{3+} proportion. In $Eu_{0.15}Tb_{1.85}$-PDC, the efficiency of energy transference (η_{ET}) from Tb^{3+} to Eu^{3+} is found to be 86.60%, which outlines Tb^{3+} as an extraordinary sensitizer in such mixed metal complex. When excited at 484 nm and observing the emission spectra at 542 nm, the time resolution spectra of $Eu_{0.15}Tb_{1.85}$-PDC were overviewed. Till 300 μs Tb^{3+} having luminescence intensity larger than that of Eu^{3+} (Figure 2.32) and an excellent green color emission is shown by this complex. Because of an increase in the rate of transference of energy from Tb^{3+} to Eu^{3+} with temperature, the green emission of Tb^{3+} due to $^5D_4 \rightarrow {}^7F_5$ transition at 542 nm diminished rapidly and that of Eu^{3+} is upgrading. After 300 μs, the emissions of both the trivalent Ln ions, that is, Tb^{3+} and Eu^{3+}, are lessening with time.

Time resolution PL spectra and fluorescence lifetime plainly display the energy transition from $Tb^{3+} \rightarrow Eu^{3+}$ ion. Furthermore, in the temperature scope of 293–333 K, $Eu_{0.05}Tb_{1.95}$-PDC acts as a self-adjusted luminous temperature sensor, with the most extreme estimations of 1.33%/K (S_a) and 1.37%/K (S_r) at 333 K. This Ln-PDC is showing amazing and remarkable optical properties; thus, it could be a luminous material having multifunctional characteristics with auspicious applications in temperature detecting, lighting and display; furthermore, time resolution imaging.

The important Ln-MOFs and their sensing properties are tabulated in Table 2.1.

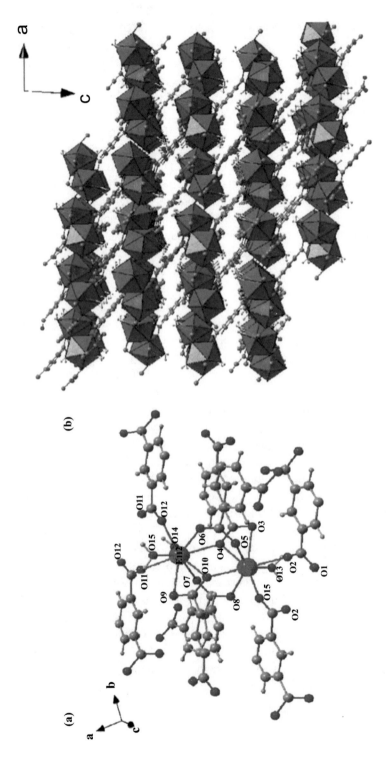

Figure 2.30: (a) Coordination environment of rare earth ion (Eu^{3+}). (b) Three-dimensional framework along *b*-axis. Reprinted with permission from [62]. Copyright (2019) American Chemical Society.

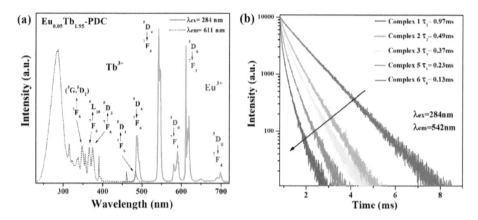

Figure 2.31: (a) The excitation and emission spectra of $Eu_{0.05}Tb_{1.95}$-PDC. (b) Fluorescence decay curves of Ln-PDC monitored at 542 nm (λ_{exc} = 284 nm). Reprinted with permission from [62]. Copyright (2019) American Chemical Society.

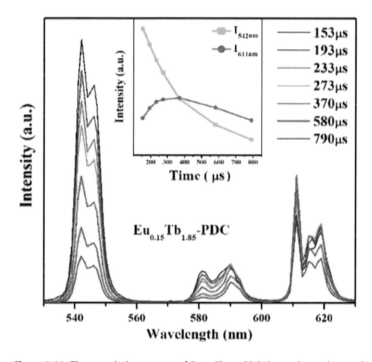

Figure 2.32: Time resolution spectra of $Eu_{0.15}Tb_{1.85}$-PDC; inset shows the graph of emission intensity versus time. Reprinted with permission from [62]. Copyright (2019) American Chemical Society.

Table 2.1: LnMOFs and their sensing properties.

Ln-MOF	Ligand	Range (K)	S_m (%/K)	T_m (K)	Ref.
$Eu_{0.0069}Tb_{0.9931}$-DMBDC	2,5-Dimethoxy-1,4-benzene dicarboxylic acid	50–200	1.15	200	[45]
$Tb_{0.9}Eu_{0.1}$-PIA	5-(Pyridine-4-yl)isophthalic acid	100–300	3.27	300	[34]
$Tb_{0.957}Eu_{0.043}$cpda	5-(4-Carboxyphenyl)-2,6-pyridinedicarboxylic acid	40–300	16.0	300	[50]

Eu$_{0.0618}$Tb$_{0.9382}$L	1,4-Bis(pyridinyl-4-carboxylato)-1,4-dimethyl benzene	25–200	0.56	200	[51]
[Eu$_{0.7}$Tb$_{0.3}$(D-cam)(Himdc)$_2$(H$_2$O)$_2$]$_3$	D-Camphoric 4,5-imidazole dicarboxylic acid	100–450	0.11	450	[52]
Tb$_{0.95}$Eu$_{0.05}$HL	5-Hydroxy-1,2,4-benzenetricarboxylic acid	4–50	31	4	[27]

(continued)

Table 2.1 (continued)

Ln-MOF	Ligand	Range (K)	S_m (%/K)	T_m (K)	Ref.
Tb$_{0.98}$Eu$_{0.02}$-OA-DSTP	2,4-(2,2':6',2''-1,4-Benzene terpyridin-4'-yl)-dicarboxylic benzenedisulfonic acid	77–275	2.4	275	[53]
Tb$_{0.98}$Eu$_{0.02}$-BDC-DSTP	Oxalic acid	125–250	2.8	225	

Ad/Tb$_{0.999}$Eu$_{0.001}$/BPDC	Biphenyl-4,4′-dicarboxylic acid / Adenine	150–250	1.27	Const. [54]
Tb$_{0.80}$Eu$_{0.20}$-BPDA	Biphenyl-3,5-dicarboxylic acid	303–328	1.39	328 [55]
Eu$_{0.0878}$Tb$_{0.9122}$L	5-(4-Tetrazol-5-yl)isophthalic acid	75–250	4.9	250 [56]

(continued)

Table 2.1 (continued)

Ln-MOF	Ligand	Range (K)	S_m (%/K)	T_m (K)	Ref.
$Eu_{0.37}Tb_{0.63}$-BTC-a	1,3,5-Benzenetricarboxylate	313–473	0.68	313	[57]
TbMOF@3%Eu-tfac	2,2′-Bipyridine,5,5′-dicarboxylate	225–375	2.59	225	[58]
TbMOF@7.3%Eu-tfac	Trifluoroacetylacetonate	200–325	1.33	325	
$Eu_{0.058}Tb_{0.942}$-BPT	Biphenyl-tricarboxylic acid	20–80 °C	1.5%/C	Const.	[59]

Eu$_{0.19}$Tb$_{0.81}$PDDI	5,5'-(Pyridine-2,5-diyl)di-isophthalic acid	313–473	0.37	473	[33]
Tb$_{0.95}$Eu$_{0.05}$FTPT	2'-Fluoro-[1,1':4'1'':4'',1'''-terphenyl]-3,3'',5,5''-tetracarboxylic acid	25–300	9.1	125	[20]
Tb$_{0.93}$Eu$_{0.07}$-BODSC	Benzophenone-3,3'-disulfonyl-4,4'-dicarboxylic acid	100–300	0.23	300	[60]

(continued)

Table 2.1 (continued)

Ln-MOF	Ligand	Range (K)	S_m (%/K)	T_m (K)	Ref.
$Eu_{0.015}Tb_{0.985}$-tatab	4,4′,4″-s-Triazine-1,3,5-triyltri-p-amino benzoic acid	230–300	-	-	[61]
$Eu_{0.0316}Tb_{0.9684}$BTPTA	4,4′,4″-(Benzene-1,3,5-triyltris(1H-pyrazole-3,1-diyl))tribenzoic acid	25–225	5.12	225	[62]

Compound	Ligand	Range (K)	Value	Ref
$Tb_{0.95}Eu_{0.05}$cpna	6-(4-Carboxyphenyl)-nicotinic acid	25–300	2.55	131	[3]
$Tb_{0.95}Eu_{0.05}$bpydc	[2,2′-Bipyridine]-5,5′-dicarboxylic acid		2.59	179	
Eu_xTb_{1-x}L	3-Bis (3-carboxyphenyl) imidazolium	320–423	1.75	423	[1]
$[(CH_3)_2NH_2]\ Eu_{0.036}Tb_{0.969}$BPTC	Biphenyl-3,3′,5,5′-tetracarboxylic acid	220–310	9.42	310	[63]
$Eu_{0.05}Tb_{1.95}$-PDC	Pyridine-3,5-dicarboxylic acid	293–333	1.37	333	[64]

2.5 Conclusion

In summary, we have concisely summarized the recent progress of the luminescent Tb/Eu-MOF thermometer. Most of the research works that are devoted to Ln-MOFs thermometry are based on the Eu^{3+}-to-Tb^{3+} emission ratio, and only a few reports investigate the emission of both an antenna (coordinated ligand) and one of these two lanthanide ions. The purposefully used organic ligand/linkers triplet excited-level energy is in the range of 22,000–27,000 cm^{-1} so that it could facilitate the energy sensitization of both trivalent Eu and Tb ions simultaneously and efficiently (the energy receiving levels are for Eu^{3+} ($^5D_1$, 19,030 cm^{-1}) and Tb^{3+} ($^5D_4$, 20,500 cm^{-1})). The future direction of lanthanide-based thermometer lies on the development of new organic antenna which not only light up lanthanides, in addition it could also participate in the sensing behavior.

References

[1] Allendorf, M. D., Bauer, C. A., Bhakta, R. K., Houk, R. J. T. (2009). Luminescent metal–organic frameworks. Chem. Soc. Rev., 38, 1330–1352.

[2] Baker, S. N., McCleskey, T. M., Baker, G. A. (2005). An ionic liquid-based optical thermometer. In ACS Symposium Series: Vol. 902. Ionic Liquids IIIB: Fundamentals, Progress, Challenges, and Opportunities, 14–171). American Chemical Society.

[3] Beeby, A. M., Clarkson, I. S., Dickins, R., Faulkner, S., Parker, D., Royle, L., . . . Woods, M. (1999). Non-radiative deactivation of the excited states of europium, terbium and ytterbium complexes by proximate energy-matched OH, NH and CH oscillators: an improved luminescence method for establishing solution hydration states. J. Chem. Soc. Perkin Trans., 2, 493–504.

[4] Binnemans, K. (2009). Lanthanide-based luminescent hybrid materials. Chem. Rev., 109, 4283–4374.

[5] Brites, C., Pereira, P., João, N., Millán, A., Amaral, V., Palacio, F., Carlos, L. (2013). Organic–Inorganic Eu^{3+}/Tb^{3+} codoped hybrid films for temperature mapping in integrated circuits. Front. Chem., 1, 9.

[6] Brites, C. D. S., Millán, A., Carlos, L. D. (2016). Chapter 281 – Lanthanides in luminescent thermometry. In: Jean-Claude, B., P. B. T.-H. on the P. and C. of R. E. Vitalij K (Eds.). Including Actinides, Vol. 49. Elsevier, 339–427.

[7] Brites, C. D. S., Lima, P. P., Silva, N. J. O., Millán, A., Amaral, V. S., Palacio, F., Carlos, L. D. (2012). Thermometry at the nanoscale. Nanoscale, 4, 4799–4829.

[8] Cadiau, A., Brites, C. D. S., Costa, P. M. F. J., Ferreira, R. A. S., Rocha, J., Carlos, L. D. (2013). Ratiometric nanothermometer based on an emissive Ln3+-organic framework. ACS Nano, 7, 7213–7218.

[9] Cui, Y., Chen, B., Qian, G. (2014). Lanthanide metal-organic frameworks for luminescent sensing and light-emitting applications. Coord. Chem. Rev., 273–274, 76–86.

[10] Cui, Y., Song, R., Yu, J., Liu, M., Wang, Z., Wu, C., . . . Qian, G. (2015). Dual-emitting MOF Dye composite for ratiometric temperature sensing. Adv. Mater., 27, 1420–1425.

[11] Cui, Y., Xu, H., Yue, Y., Guo, Z., Yu, J., Chen, Z., . . . Chen, B. (2012). A luminescent mixed-lanthanide metal–organic framework thermometer. J. Am. Chem. Soc., 134, 3979–3982.
[12] Cui, Y., Yue, Y., Qian, G., Chen, B. (2012). Luminescent functional metal–organic frameworks. Chem. Rev., 112, 1126–1162.
[13] Cui, Y., Zhang, J., He, H., Qian, G. (2018). Photonic functional metal–organic frameworks. Chem. Soc. Rev., 47, 5740–5785.
[14] Cui, Y., Zhu, F., Chen, B., Qian, G. (2015). Metal–organic frameworks for luminescence thermometry. Chem. Commun., 51, 7420–7431.
[15] Cui, Y., Zou, W., Song, R., Yu, J., Zhang, W., Yang, Y., Qian, G. (2014). A ratiometric and colorimetric luminescent thermometer over a wide temperature range based on a lanthanide coordination polymer. Chem. Commun., 50, 719–721.
[16] D'Vries, R. F., Álvarez-García, S., Snejko, N., Bausá, L. E., Gutiérrez-Puebla, E., De Andrés, A., Monge, M. Á. (2013). Multimetal rare earth MOFs for lighting and thermometry: tailoring color and optimal temperature range through enhanced disulfobenzoic triplet phosphorescence. J. Mater. Chem. C, 1, 6316–6324.
[17] Døssing, A. (2005). Luminescence from lanthanide(3+) ions in solution. Eur. J. Inorg. Chem., 2005, 1425–1434.
[18] Dou, Z., Yu, J., Cui, Y., Yang, Y., Wang, Z., Yang, D., Qian, G. (2014). Luminescent metal–organic framework films as highly sensitive and fast-response oxygen sensors. J. Am. Chem. Soc., 136, 5527–5530.
[19] Han, Y.-H., Tian, C.-B., Li, Q.-H., Du, S.-W. (2014). Highly chemical and thermally stable luminescent Eu$_x$Tb$_{1-x}$ MOF materials for broad-range pH and temperature sensors. J. Mater. Chem. C, 2, 8065–8070.
[20] Jaque, D., Vetrone, F. (2012). Luminescence nanothermometry. Nanoscale, 4, 4301–4326.
[21] Kaczmarek, A. M. (2018). Eu^{3+}/Tb^{3+} and Dy^{3+} POM@MOFs and 2D coordination polymers based on pyridine-2,6-dicarboxylic acid for ratiometric optical temperature sensing. J. Mater. Chem. C, 6, 5916–5925.
[22] Kaczmarek, A. M., Liu, -Y.-Y., Wang, C., Laforce, B., Vincze, L., Van Der Voort, P., Van Deun, R. (2017). Grafting of a Eu3+-tfac complex on to a Tb^{3+}-metal organic framework for use as a ratiometric thermometer. Dalton Trans., 46, 12717–12723.
[23] Khalil, G. E., Lau, K., Phelan, G. D., Carlson, B., Gouterman, M., Callis, J. B., Dalton, L. R. (2003). Europium beta-diketonate temperature sensors: effects of ligands, matrix, and concentration. Rev. Sci. Instrum., 75, 192–206.
[24] Kim, K., Jeong, W., Lee, W., Reddy, P. (2012). Ultra-High Vacuum Scanning Thermal Microscopy for Nanometer Resolution Quantitative Thermometry. ACS Nano, 6, 4248–4257.
[25] Latva, M., Takalo, H., Mukkala, V.-M., Matachescu, C., Rodríguez-Ubis, J. C., Kankare, J. (1997). Correlation between the lowest triplet state energy level of the ligand and lanthanide (III) luminescence quantum yield. J. Lumin., 75, 149–169.
[26] Li, L., Lin, R.-B., Krishna, R., Li, H., Xiang, S., Wu, H., . . . Chen, B. (2018). Ethane/ethylene separation in a metal-organic framework with iron-peroxo sites. Science, 362, 443 LP –446.
[27] Li, Y., Zhang, S., Song, D. (2013). A luminescent metal–organic framework as a turn-on sensor for DMF vapor. Angew. Chem. Int. Ed., 52, 710–713.
[28] Liu, G., Di Yuan, Y., Wang, J., Cheng, Y., Peh, S. B., Wang, Y., . . . Zhao, D. (2018). Process-tracing study on the postassembly modification of highly stable zirconium metal–organic cages. J. Am. Chem. Soc., 140, 6231–6234.
[29] Liu, X., Akerboom, S., Jong De, M., Mutikainen, I., Tanase, S., Meijerink, A., Bouwman, E. (2015). Mixed-Lanthanoid Metal–Organic Framework for Ratiometric Cryogenic Temperature Sensing. Inorg. Chem., 54, 11323–11329.

[30] Lustig, W. P., Mukherjee, S., Rudd, N. D., Desai, A. V., Li, J., Ghosh, S. K. (2017). Metal-organic frameworks: functional luminescent and photonic materials for sensing applications. Chem. Soc. Rev., 46, 3242–3285.

[31] McLaurin, E. J., Bradshaw, L. R., Gamelin, D. R. (2013). Dual-emitting nanoscale temperature sensors. Chem. Mater., 25, 1283–1292.

[32] Meng, X., Song, S.-Y., Song, X.-Z., Zhu, M., Zhao, S.-N., Wu, L.-L., Zhang, H.-J. (2014). A Eu/Tb-codoped coordination polymer luminescent thermometer. Inorg. Chem. Front., 1, 757–760.

[33] Miyata, K., Konno, Y., Nakanishi, T., Kobayashi, A., Kato, M., Fushimi, K., Hasegawa, Y. (2013). Chameleon luminophore for sensing temperatures: control of metal-to-metal and energy back transfer in lanthanide coordination polymers. Angew. Chem. Int. Ed., 52, 6413–6416.

[34] Pan, Y., Su, H.-Q., Zhou, E.-L., Yin, H.-Z., Shao, K.-Z., Su, Z.-M. (2019). A stable mixed lanthanide metal–organic framework for highly sensitive thermometry. Dalton Trans., 48, 3723–3729.

[35] Parker, D. (2000). Luminescent lanthanide sensors for pH, pO2 and selected anions. Coord. Chem. Rev., 205, 109–130.

[36] Rao, X., Song, T., Gao, J., Cui, Y., Yang, Y., Wu, C., . . . Qian, G. (2013). A highly sensitive mixed lanthanide metal–organic framework self-calibrated luminescent thermometer. J. Am. Chem. Soc., 135, 15559–15564.

[37] Ren, M., Brites, C. D. S., Bao, -S.-S., Ferreira, R. A. S., Zheng, L.-M., Carlos, L. D. (2015). A cryogenic luminescent ratiometric thermometer based on a lanthanide phosphonate dimer. J. Mater. Chem. C, 3, 8480–8484.

[38] Rocha, J., Brites, C. D. S., Carlos, L. D. (2016). Lanthanide organic framework luminescent thermometers. Chem. Eur. J., 22, 14782–14795.

[39] Rocha, J., Carlos, L. D., Paz, F. A. A., Ananias, D. (2011). Luminescent multifunctional lanthanides-based metal–organic frameworks. Chem. Soc. Rev., 40, 926–940.

[40] Shen, X., Lu, Y., Yan, B. (2015). Lanthanide complex hybrid system for fluorescent sensing as thermometer. Eur. J. Inorg. Chem., 2015, 916–919.

[41] Tang, Y.-Z., Yang, Y.-M., Wang, X.-W., Zhang, Q., Wen, H.-R. (2011). Synthesis, structure and XPS of a novel two-dimensional CuII–EuIII heterometallic–organic framework. Inorg. Chem. Commun., 14, 613–617.

[42] Vetrone, F., Naccache, R., Zamarrón, A., Juarranz De La Fuente, A., Sanz-Rodríguez, F., Martinez Maestro, L., . . . Capobianco, J. A. (2010). Temperature Sensing Using Fluorescent Nanothermometers. ACS Nano, 4, 3254–3258.

[43] Wade, S. A., Collins, S. F., Baxter, G. W. (2003). Fluorescence intensity ratio technique for optical fiber point temperature sensing. J. Appl. Phys., 94, 4743–4756.

[44] Wang, H., Zhao, D., Cui, Y., Yang, Y., Qian, G. (2017). A Eu/Tb-mixed MOF for luminescent high-temperature sensing. J Solid State Chem., 246, 341–345.

[45] Wang, X., Wolfbeis, O. S., Meier, R. J. (2013). Luminescent probes and sensors for temperature. Chem. Soc. Rev., 42, 7834–7869.

[46] Wang, Z., Ananias, D., Carné-Sánchez, A., Brites, C. D. S., Imaz, I., Maspoch, D., . . . Carlos, L. D. (2015). Lanthanide–organic framework nanothermometers prepared by spray-drying. Adv. Funct. Mater., 25, 2824–2830.

[47] Wei, Y., Sa, R., Li, Q., Wu, K. (2015). Highly stable and sensitive LnMOF ratiometric thermometers constructed with mixed ligands. Dalton Trans., 44, 3067–3074.

[48] Wu, -L.-L., Zhao, J., Wang, H., Wang, J. (2016). A lanthanide(iii) metal–organic framework exhibiting ratiometric luminescent temperature sensing and tunable white light emission. Cryst. Eng. Comm., 18, 4268–4271.

[49] Yang, Y., Huang, H., Wang, Y., Qiu, F., Feng, Y., Song, X., . . . Liu, W. (2018). A family of mixed-lanthanide metal–organic framework thermometers in a wide temperature range. Dalton Trans., 47, 13384–13390.

[50] Yang, Y., Wang, Y., Feng, Y., Song, X., Cao, C., Zhang, G., Liu, W. (2020). Three isostructural Eu^{3+}/Tb^{3+} co-doped MOFs for wide-range ratiometric temperature sensing. Talanta, 208, 120354.

[51] Yao, J., Zhao, Y.-W., Zhang, X.-M. (2018). Breathing europium–terbium co-doped luminescent MOF as a broad-range ratiometric thermometer with a contrasting temperature–intensity relationship. ACS Omega, 3, 5754–5760.

[52] Yi, F.-Y., Chen, D., Wu, M.-K., Han, L., Jiang, H.-L. (2016). Chemical sensors based on metal–organic frameworks. ChemPlusChem, 81, 675–690.

[53] Zhang, L., Xie, Y., Xia, T., Cui, Y., Yang, Y., Qian, G. (2018). A highly sensitive luminescent metal–organic framework thermometer for physiological temperature sensing. J. Rare Earths, 36, 561–566.

[54] Zhang, -W.-W., Wang, Y.-L., Liu, Q., Liu, Q.-Y. (2018). Lanthanide-benzophenone-3,3′-disulfonyl-4,4′-dicarboxylate Frameworks: temperature and 1-Hydroxypyren luminescence sensing and proton conduction. Inorg. Chem., 57, 7805–7814.

[55] Zhao, D., Cui, Y., Yang, Y., Qian, G. (2016). Sensing-functional luminescent metal–organic frameworks. Cryst. Eng. Comm., 18, 3746–3759.

[56] Zhao, D., Rao, X., Yu, J., Cui, Y., Yang, Y., Qian, G. (2015). Design and synthesis of an MOF thermometer with high sensitivity in the physiological temperature range. Inorg. Chem., 54, 11193–11199.

[57] Zhao, D., Wang, H., Qian, G. (2018). Synthesis, structure and temperature sensing of a lanthanide-organic framework constructed from a pyridine-containing tetracarboxylic acid ligand. Cryst. Eng. Comm., 20, 7395–7400.

[58] Zhao, D., Yue, D., Jiang, K., Zhang, L., Li, C., Qian, G. (2019). Isostructural Tb^{3+}/Eu^{3+} Co-doped metal–organic framework based on pyridine-containing dicarboxylate ligands for ratiometric luminescence temperature sensing. Inorg. Chem., 58, 2637–2644.

[59] Zhao, D., Yue, D., Zhang, L., Jiang, K., Qian, G. (2018). Cryogenic Luminescent Tb/Eu-MOF Thermometer Based on a Fluorine-Modified Tetracarboxylate Ligand. Inorg. Chem., 57, 12596–12602.

[60] Zhao, S.-N., Li, L.-J., Song, X.-Z., Zhu, M., Hao, Z.-M., Meng, X., . . . Zhang, H.-J. (2015). Lanthanide ion codoped emitters for tailoring emission trajectory and temperature sensing. Adv. Funct. Mater., 25, 1463–1469.

[61] Zhou, H.-C., Joe,, Kitagawa, S. (2014). Metal–Organic Frameworks (MOFs). Chem. Soc. Rev., 43, 5415–5418.

[62] Zhou, X., Wang, H., Jiang, S., Xiang, G., Tang, X., Luo, X., . . . Zhou, X. (2019). Multifunctional Luminescent Material Eu(III) and Tb(III) Complexes with Pyridine-3,5-Dicarboxylic Acid Linker: crystal Structures, Tunable Emission, Energy Transfer, and Temperature Sensing. Inorg. Chem., 58, 3780–3788.

[63] Zhou, Y., Yan, B., Lei, F. (2014). Postsynthetic lanthanide functionalization of nanosized metal–organic frameworks for highly sensitive ratiometric luminescent thermometry. Chem. Commun., 50, 15235–15238.

[64] Zhu, Y., Xia, T., Zhang, Q., Cui, Y., Yang, Y., Qian, G. (2018). A Eu/Tb mixed lanthanide coordination polymer with rare 2D thick layers: synthesis, characterization and ratiometric temperature sensing. J Solid State Chem, 259, 98–103.

I. V. García-Amaya, L. Castro-Arce, G. A. Limon-Reynosa,
J. Manzanares-Martinez, M. C. Acosta-Enríque, Ma. E. Zayas
Chapter 3
Redshift in Tellurite Glasses Doped with Eu^{3+} due to Heat Treatment

Abstract: Oxide glasses were prepared from $17ZnO \bullet 32CdO \bullet 51TeO_2$ composition with varying amounts of $Eu(NO_3)_3 \cdot 6H_2O$ (0–1.5 mol%), at 1,000 °C by the conventional melt-quenching method. After the cooling process, glasses have to undergo heat treatments at 300 and 420 °C for 5 h. X-ray diffraction results show an amorphous structure except for the heat treatment at 420 °C. We have found that the nanometric crystalline phases are more abundant with a low europium concentration. And also the crystalline phase of $ZnTeO_3$ dominates in all glass samples with a crystallite size of around 36 nm. On the other hand, the density of glasses was determined, and it is observed that it is higher for the type of as-cast glasses, in the range of 5.70–5.78.

The spectra of the doped glasses reveal absorption peaks of the Eu^{3+} ion at 350, 362, 377, 382, 395, 415 and 465 nm associated with the transitions $^7F_0 \rightarrow {}^5L_7$, $^7F_0 \rightarrow {}^5D_4$, $^7F_0 \rightarrow {}^5G_4$, $^7F_0 \rightarrow {}^5G_2$, $^7F_0 \rightarrow {}^5L_6$, $^7F_1 \rightarrow {}^5D_3$ and $^7F_0 \rightarrow {}^5D_2$, respectively. Photoluminescence spectra obtained with an excitation wavelength of 392 nm show bands related to the emissions due to the Eu^{3+} transitions. The intensity of radius (R) increases with heat treatments, suggesting that the asymmetry around the europium ions increases. The chromaticity coordinate and color purity were calculated, and the results indicate a redshift. The decay time is very similar in all samples and does not vary with the concentration of europium.

Keywords: tellurite glass, Eu^{3+} emission, chromaticity coordinates, decay time

---

I. V. García-Amaya, Programa de Ingeniería en Geociencias, Universidad Estatal de Sonora, Unidad Hermosillo, Ley Federal del Trabajo S/N, Colonia Apolo, Hermosillo, Sonora, C.P. 83100, México; Unidad Magdalena, Guerrero Sur 917, Colonia San Martin, Magdalena de Kino Sonora, C.P. 84160
L. Castro-Arce, Departamento de Física, Matemáticas e Ingeniera, Universidad de Sonora, Unidad Sur Navojoa, Sonora, México
G. A. Limon-Reynosa, Departamento de Ciencias Quimico Bilogicas y Agropecuarias, Universidad de Sonora, Unidad Sur Navojoa, Sonora, México
J. Manzanares-Martinez, M. C. Acosta-Enríque, Ma. E. Zayas, Departamento de Investigación en Física, Universidad de Sonora, Blvd. Transversal y Rosales S/N, Colonia Centro, Hermosillo, Sonora, C.P. 83000, México

https://doi.org/10.1515/9783110676457-003

3.1 Introduction

In the luminescent materials study, it is very common to focus on solid materials like crystals. However, there are amorphous materials that when doped with rare earth ions luminescence is presented among them such as, glasses, plastics, and polymers. RE^{3+}-doped glasses have potential applications on solid-state devices, visible and infrared (IR) lasers, fiber amplifiers, light-emitting diodes (LEDs), and 3D displays [5, 20, 32].

Among the RE ions, europium is an essential optical activator that plays a fundamental role in applications on solid-state lasers, optical communication systems, color display devices, sensors, and LEDs [24]. An exciting characteristic of europium is that it can exist in various oxidation states, and in both cases, it generates a strong luminescence in the visible region, that is, blue or green emission in the near-UV region for the Eu^{2+} state and red emission for the Eu^{3+} state [2].

Trivalent europium ions exhibit a narrow emission band in the reddish-orange or bright red region of the electromagnetic spectrum (613 nm) and long lifetimes [2, 24, 35]. The non-degenerate nature of the (excited) $^5D_0$ and the $^7F_0$ (ground) states of Eu^{3+} ions provide valuable information on symmetry and the host matrix coordination environment [15]. Therefore, Eu^{3+} ions are used on spectroscopic probes for the study of symmetry and heterogeneity for different receptor matrices [33].

However, in general, the Eu^{3+} oxidation state cannot be efficiently excited by near-UV light since its excitation peaks are in the region due to f–f forbidden parity transitions [2]. The radiative properties of RE^{3+} ions depend on the local structure and the chemical composition of the host matrices [24]. Therefore, to obtain efficient luminescence, suitable host materials are needed. In this sense, vitreous materials are usually an excellent choice as luminescence materials due to their characteristic structural disorder, their diverse compositions and their easy production [2].

Compared to other phosphors, Eu^{3+} doped glasses are an alternative since they have the advantage of presenting a homogeneous light emission. They reduce the fabrication cost, the manufacturing processes are more straightforward and they have greater thermal stability [38].

Tellurite glasses are known for their low phonon energy, high refractive index and overall high optical basicity, enhancing the reduction of doped elements [13, 28]. Tellurite glasses are also known for their low glass transition temperature and low melting temperatures.

Tellurium oxide is a particular case of a glass former; the structural unit constitutes that the network has an octahedral coordination nature, which deviates from the classical structural model [8].

On the other hand, glasses containing heavy metal oxides as network modifiers have a high refractive index, high transmission in the IR region and high density, so they are used for technological applications [18, 40]. In particular, CdO, as a network modifier, in addition to stabilizing the vitreous network, improves the electrical properties of glasses. However, at high CdO contents, the Cd^{2+} ion can alter the glass structure, since its cations can enter the structure and act as network modifiers or glass formers, as a consequence of its high polarizability" [22, 23, 27].

The function of zinc oxide when introduced as a component of a glassy matrix that contains TeO_2 in its chemical composition is of relevant importance. Because it can change the coordination from TeO_4 units to TeO_3 units, the TeO_{3+1} polyhedron appearing in an intermediate way is unstable and stabilizes in TeO_3. But it is worth mentioning that TeO_4 forms trigonal bipyramids, and some of them can be asymmetric containing a $Te-O_{ax}$ bond, an elongated axial one for low ZnO concentrations, and the other one shortens, producing TeO_{3+1} polyhedra [34]. But it has been seen that when the vitreous reticulum is saturated with ZnO, TeO_{3+1} is affected to the degree that oxygen with elongated bond is detached, and this polyhedron is stabilized in a trigonal pyramid characteristic of TeO_3 groups, causing that one of the remaining oxygen atoms forms a double terminal bond Te = O [34].

Recent structural studies demonstrated that zinc oxide plays the role of a modifier and a network former, leading to changes in the physical and optical properties of tellurium glasses [9, 21]. Likewise, zinc provides good chemical durability and low glass transition temperature since it reduces the melting point of glass [5].

Zayas et al. [41] studied for the first time the glass-forming zone of the ternary $ZnO-CdO-TeO_2$ system using the conventional powder fusion method. This zone covers a very narrow portion of the ternary composition, and glass formation occurs at high concentrations of TeO_2, giving rise to transparent, opaque and partially crystallized glasses. The amount of TeO_2 greater than 50 wt% produces transparent glasses. On the other hand, for high concentrations of CdO, the glasses become opaque due to the crystallization of the $CdTeO_3$ phase.

The properties of $ZnO-CdO-TeO_2$ vitreous system with the incorporation of RE compounds allow different amounts of CdO and TeO_2 concentrations. Ruvalcaba-Cornejo et al. [29] reported the optical properties of the vitreous matrix doped with RE compounds $YbBr_3$ and Tb_4O_7, finding the formation of the $CdTeO_3$ phase with the presence of Tb^{3+} ions". In the same manner, Ruvalcaba-Cornejo et al. [31] reported doping with $NdCl_3$ for finding amorphous glasses for all compositions. Finally, Ruvalcaba-Cornejo et al. [30] studied this system adding $EuCl_3$, finding that the less amount of TeO_2 used in the glass, favours the develop of the $CdTe_2O_5$ crystalline phase. Also, finding that the crystal size it is around 10 nm. On the other hand, the effect of europium doping on the photoluminescence (PL) of $ZnO-CdO-TeO_2$ glasses was studied [11], and scanning electron microscopic (SEM) observations revealed the formation of zinc aluminate spinel. The "Raman spectra evidenced a band located at 1,556 cm^{-1} that is related to interstitial molecular oxygen in the glass matrix.

García-Amaya et al. [10] studied glasses of the $17ZnO \cdot 32CdO \cdot 51TeO_2$ system doped with Eu_2O_3 (0–0.75 mol per 100 mol of oxides). The samples were exposed to thermal treatment in a laboratory oven at 420 °C. X-ray diffraction (XRD) analysis of the thermal-treated samples showed crystalline phases of a variety of metallic oxides, and the intensities of which decreased as the Eu doping amount in the samples increased. The most substantial peak corresponded to $ZnTeO_3$ with a crystallite size of about 36 nm in all samples. The observations by different structural techniques suggest that Eu_2O_3-induced structural changes led to the formation of molecular oxygen in the interstitial network".

The purpose of this research is to study the influence of the Eu^{3+} ion when its concentration varied in the vitreous matrix, likewise, to promote the development of nanocrystalline phases through heat treatments, to study PL and to improve optical effects in order to suggest its possible applications in lasers and television screens.

3.2 Experimental Procedure

The content of metallic oxides in the glass matrix was maintained constant as indicated in the chemical formula "17ZnO•32CdO•51TeO$_2$" in mol% (Table 3.1).[1] The batch mixtures were processed by the melt-quenching method in high alumina crucibles at 1,000 °C for 30 min in a Thermolyne 46100 furnace. Further, the glasses were quickly transferred into an annealing Thermolyne 48000 furnace at 250 °C for 20 min, and then cooled down slowly to room temperature. Based on the differential scanning calorimetry results and $ZnO-TeO_2$ and $CdO-TeO_2$ phase diagrams [1, 35], temperatures of 300 and 420 °C were selected to carry out the heat treatment of as-cast glasses. The samples were maintained at that temperature for 5 h in a Thermolyne 48000 furnace. A small amount of the glassy material was crushed and milled down to a particle size under 30 μm in an agate mortar to conduct XRD analyses in a Bruker D8 advanced diffractometer. For that purpose, we analyzed the Cu-Kα radiation ($\lambda = 1.5418$ Å) in the 2θ range of 15–65°. The density of the glasses was determined by the difference in weight of the material and the apparent weight in immersion using Archimedes' principle:

$$\rho = \frac{W_a}{(W_a - W_b)\rho_x}$$

where W_a is the weight of the sample in air and W_b is the weight of the sample when immersed in distilled water of density $\rho_x = 0.99681$ g/cm^3 at a temperature of 26 °C.

Density values were obtained using an OHAUS balance (model GA110) with an accuracy of 0.0001 g.

The optical band gap was calculated assuming an indirect band gap using the Tauc method, since these materials are amorphous [2]. Finally, PL was acquired in a spectrofluorometer using either a 450 W ozone-free Xe lamp, or a 325 nm He–Cd laser, coupled with a Horiba Triax 320 excitation and an emission iHR320 monochromator.

Table 3.1: Sample ID and nominal composition of the 17ZnO·32CdO·51TeO$_2$ samples. Taken from García-Amaya et al. [10], p. 50.

Sample ID	Dopant concentration, mol% Eu(NO$_3$)$_3$ · 6H$_2$O/100 mol of oxides
ZCT	0
ZCT-0.3	0.3
ZCT-0.6	0.6
ZCT-0.9	0.9
ZCT-1.2	1.2
ZCT-1.5	1.5

3.3 Results and Discussion

The glasses obtained are located in the Gibbs triangle (Figure 3.1). The physical characteristics of the three series are given in Table 3.2. We appreciate to make an eye that the as-cast glasses are light green, transparent and compact. In the T300 series, the glasses are colorless, brighter and transparent. In the T420 series, all glasses are opaque but have two colors: ZCT, ZCT-1.2, ZCT-1.5 are whitish and ZCT-0.3, ZCT-0.6 and ZCT-0.9 are greenish.

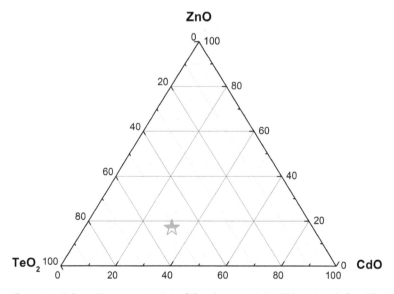

Figure 3.1: Schematic representation of the glass matrix in Gibbs' triangle (modified from [41]).

Table 3.2: Aspects of the glasses obtained.

Sample ID	As-cast	T300	T420
ZCT			
ZCT-0.3			
ZCT-0.6			
ZCT-0.9			
ZCT-1.2			
ZCT-1.5			

The information related to the final composition of the base glasses analyzed by SEM/energy-dispersive X-ray spectroscopy was published in [11]. With the addition of $Eu(NO_3)_3 \cdot 6H_2O$, Cd and Te concentrations increase slightly, while the content of O and Zn decreases compared to undoped glass (ZCT). Among the doped glasses, there are no significant variations in Cd, Zn and Te concentration, and the increase in Eu content is as expected. Also, in all glasses a slight amount of aluminum was detected, probably due to corrosion of the high alumina crucible used during melting.

3.3.1 DRX

Figure 3.2a[2] shows the XRD results of the glasses without heat treatment (as-cast), which is typical of amorphous materials in the systems. After heat treatment at 300 °C (T300), the glasses keep their amorphous structure, showing slight changes in the band intensity, which could indicate a tendency of the system to order in short range (Figure 3.2b). At the same time, heat treatment at 420 °C (T420) (Figure 2c[3]) shows sharp diffraction peaks overlapped with the band from the glass matrix, where glasses with concentrations of $Eu(NO_3)_3 \cdot 6H_2O$ between 0 and 0.6 mol% present development of crystalline phases, and the most dominant being the $ZnTeO_3$ phase, as shown in Table 3.3. Crystallite size and discussion of diffraction patterns can be found in [10].

Table 3.3: Crystalline phases detected in the XRD diffractograms of all T420 series glasses.

No.	Crystalline phase	Crystal size (nm)	Reference
1	$ZnTeO_3$	36	PDF#44-0240
2	Te	45	PDF#02-0511
3	$CdTeO_3$	31	PDF#36-0890
4	$Zn_2Te_3O_8$	38	PDF#44-0241
5	ZnO	27	PDF#03-0752
6	CdO_2	24	PDF#78-1125

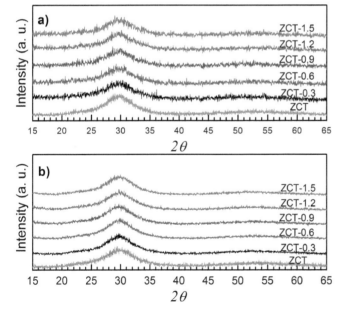

Figure 3.2: X-ray diffraction patterns of glasses. a) As-cast, b) T300 series and c). T420 series. Figures 3.2a and 3.3c was taken from García-Amaya et al. [10], p. 51.

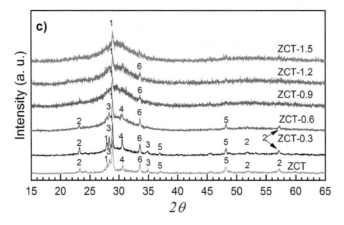

Figure 3.2 (continued)

3.3.2 Density

Density is a physical property of materials. When the density suffers fluctuations in its values, structural changes are noted that occur in the glasses under study. According to Table 3.4, the density values of the glasses as-cast vary in the 5.77–5.70 g/cm^3 range, observing a linear decrease of the density due to the change in the coordination of ZnO_6 to ZnO_4 and also the Zn^{2+} ions produce a break of the Te–O–Te bonds (Figure 3.3). While for europium oxide concentrations of 1.2–1.5, the density value increases because ZnO favors the transformation from TeO_3 to TeO_4 [11].

For the T300 series glasses, a nonlinear density behavior is observed for all europium oxide concentrations of 0.3–1.5. This is most probably due to the structural rearrangement that occurs in the vitreous reticulum due to the effects of heat treatment.

For the T420 series glasses, the density decreases when the europium oxide varies from 0.3 to 0.6, due to the effects of the transformation of the TeO_{3+1} polyhedron to TeO_3. While for europium oxide concentration of 0.9–1.2, the density increases because TeO_3 is transformed into TeO_4, and for the europium oxide ratio of 1.5, the density decreases due to the presence of Te–O–Te and O–Te–O bonds [11].

3.3.3 Optical Absorption

The absorption spectra of the glass matrix of ZCT doped with europium ions, recorded in the UV–Vis region, are shown in Figure 3.4. The spectra of the doped glasses reveal absorption peaks of the Eu^{3+} ion at 350, 362, 377, 382, 395, 415 and 465 nm associated with the transitions $^7F_0 \rightarrow {}^5L_7$, $^7F_0 \rightarrow {}^5D_4$, $^7F_0 \rightarrow {}^5G_4$, $^7F_0 \rightarrow {}^5G_2$, $^7F_0 \rightarrow {}^5L_6$, $^7F_1 \rightarrow {}^5D_3$ and $^7F_0 \rightarrow {}^5D_2$, respectively [7, 15, 17, 25, 33, 36, 37].

Table 3.4: Density of the glass series studied in the ZnO–CdO–TeO$_2$: Eu^{3+} system.

Samples	As-cast	T300	T420
		(g/cm^3)	
ZCT	5.83	5.60	5.61
ZCT-0.3	5.77	5.58	5.52
ZCT-0.6	5.73	5.65	5.50
ZCT-0.9	5.70	5.48	5.73
ZCT-1.2	5.73	5.49	5.78
ZCT-1.5	5.78	5.25	5.68

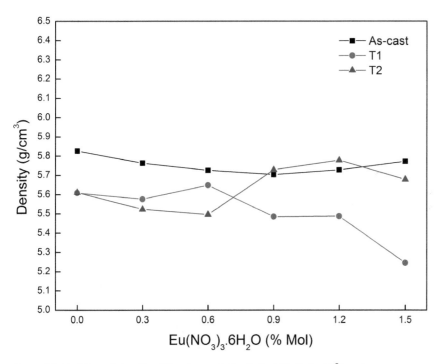

Figure 3.3: Variation of density of the glasses in the ZnO–CdO–TeO$_2$: Eu^{3+} system.

Figure 3.4a and b presents a general behavior of the peaks which increase their intensity with a higher content of the dopant, which suggests that the higher concentration of europium increases the density of defects in the material [26]. Meanwhile, the ZCT glasses are different from those described above, where the matrix of the as-cast shows a jump, produced by the equipment, around 380 nm; on the other hand, T300 series glasses exhibit two small bands around 352 and 357 nm, which may be due to defects produced in the glassy matrix.

We observe that the T420 series glasses (Figure 4c) are obtained by using diffuse reflectance, since they are opaque and translucent glasses; therefore, the absorption spectra are very different from those of the other series. In this series, the absorption peaks of europium have a greater amplitude. Likewise, they do not vary linearly with the concentration of the dopant. The absorption spectrum of the matrix changes the absorption edge at ~ 420 nm.

3.3.4 Band Gap

The optical band gap energy values were determined by the intersection of the tangent line (extrapolated part of the linear region) using the Tauc method ($(\alpha h\nu)^{1/2}$ vs $h\nu$)) with the abscissa axis [12]. Figure 3.5a–c shows the Tauc diagram as a function of photon energy ($h\nu$), and the band gap energy values are given in Table 3.5.

The doped glasses have a reduction of the band gap energy by increasing the concentration of europium. In T300 series, the band gap energy of the matrix is lower than that of doped glasses. In the T420 series, the matrix tends to have the same behavior of the T300 series, and the ZCT-0.3 and ZCT-0.6 glasses increase gradually; in ZCT-0.9, the band gap energy decreases; in ZCT-1.2 and ZCT-1.5, it increases again.

In general, the band gap of ZCT glasses is lower than the doped glasses; the T300 series has the smallest band gap, followed by as-cast; the glasses with the highest band gap are the T420 series.

3.3.5 Photoluminescence

The luminescent properties of Eu^{3+} ion-doped ZCT system glasses determine that the environment influences the luminescence intensity of the europium ions. The excitation spectra of the as-cast series glasses monitored at λ_{em} = 613 nm (transition $^5D_0 \rightarrow ^7F_2$ from Eu^{3+}) were published in [11]. The characteristic bands are originated by the 4f–4f transitions of Eu^{3+} from its $^7F_0$ basal state to the excitation levels: $^5D_4$ (361 nm), $^5L_6$ (393 nm), $^5D_3$ (415 nm), $^5D_2$ (464 nm) and $^5D_1$ (531 nm) [1, 20]. The intensity of the peaks at 393 and 464 nm is very similar and is significantly higher than the other bands. The bands at 393 and 531 nm show changes in intensity at different

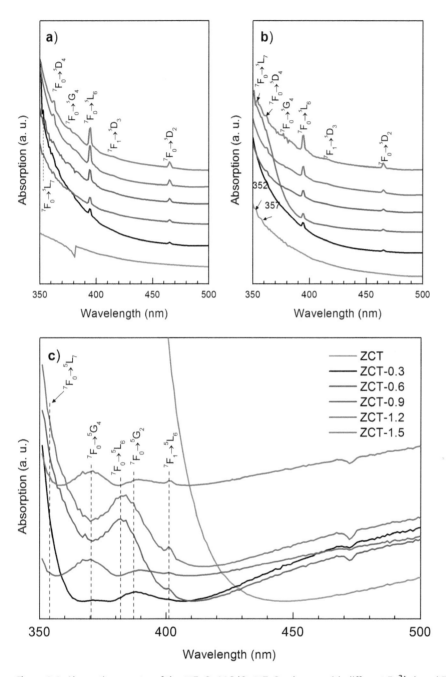

Figure 3.4: Absorption spectra of the 17ZnO•32CdO•51TeO$_2$ glasses with different Eu^{3+}-doped ZCT glasses: (a) as-cast, (b) T300 and (c) T420 series.

Table 3.5: Optical band gap values spectra of the 17ZnO•32CdO•51TeO$_2$ glasses with different Eu^{3+}-doped ZCT glasses: (a) as-cast, (b) T300 and (c) T420, series.

Samples	As-cast	T300	T420
	(eV)	(eV)	(eV)
ZCT	2.66	3.15	3.08
ZCT-0.3	3.48	3.33	3.46
ZCT-0.6	3.35	3.34	3.48
ZCT-0.9	3.29	3.23	3.41
ZCT-1.2	3.28	3.22	3.43
ZCT-1.5	3.25	3.44	3.49

concentrations of europium. The excitation spectra of the T300 and T420 series are similar.

Eu^{3+} ion PL spectra on ZCT glasses were obtained with an excitation wavelength of level $^5L_6$ (λ_{exc} = 392 nm). The characteristic emission bands correspond to the transition $^5D_0 \rightarrow {}^7F_j$ (where j = 2, 1). All PL spectra (Figure 3.6a–c,[4]) have a total of eight bands, which are related to europium ion transitions: $^5D_2 \rightarrow {}^7F_3$ (511 nm), $^5D_1 \rightarrow {}^7F_1$ (536 nm), $^5D_1 \rightarrow {}^7F_2$ (554 nm), $^5D_0 \rightarrow {}^7F_0$ (579 nm), $^5D_0 \rightarrow {}^7F_1$ (592 nm), $^5D_0 \rightarrow {}^7F_2$ (613 nm), $^5D_0 \rightarrow {}^7F_3$ (652 nm) and $^5D_0 \rightarrow {}^7F_4$ (701 nm). These spectra are similar to those obtained in other Eu^{3+}-doped glass systems [6, 15, 17, 32, 33].

It is known that the $^5D_0 \rightarrow {}^7F_1$ transition is a magnetic dipole and is independent of local symmetry, while the transition $^5D_0 \rightarrow {}^7F_2$ is a characteristic electrical dipole transition and is known to be a hypersensitive transition because it is strongly influenced by the Eu^{3+} ion environment [1, 2, 4, 36]. For this reason, the intensity of the luminescent radius (R) of the transitions $^5D_0 \rightarrow {}^7F_2$ (red)/$^5D_0 \rightarrow {}^7F_1$ (orange) is defined as the ratio of red to orange fluorescence intensity and provides us with information about the local symmetry of the Eu^{3+} ions, indicating the covalent/ionic bonding strength between the ions and the surrounding ligands [1, 3, 4, 32]. It means that the transition $^5D_0 \rightarrow {}^7F_2$ is much stronger than the transition $^5D_0 \rightarrow {}^7F_1$. It suggests that the Eu^{3+} ion is located in a distorted (asymmetric) cationic environment [42]. Likewise, the appearance of the nondegenerated transition $^5D_0 \rightarrow {}^7F_0$ (579 nm) and the division of the transition $^5D_0 \rightarrow {}^7F_1$ (592 nm) into three components confirms the minor symmetry of the Eu^{3+} sites such as orthorhombic, monoclinic or triclinic in the vitreous reticulum [4, 14]. For better comparison (see Figure 3.6), the PL spectra were normalized at a wavelength of 590 nm (transition $^5D_0 \rightarrow {}^7F_1$), as it is independent of the glass matrix [1, 4].

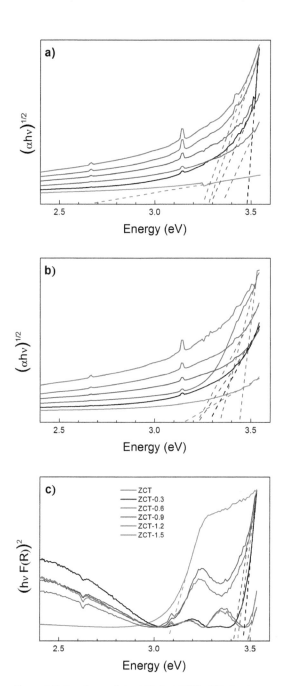

Figure 3.5: Band gap of the 17ZnO•32CdO•51TeO$_2$ glasses with different Eu^{3+}-doped ZCT glasses: (a) as-cast, (b) T300 and (c) T420 series.

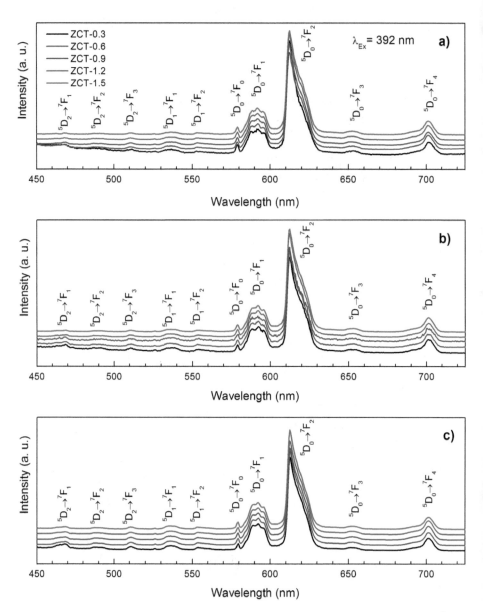

Figure 3.6: Photoluminescence spectrum of the 17ZnO·32CdO·51TeO$_2$ glasses with different Eu^{3+} doped obtained at λ_{ex} = 294 nm, a) As cast, b) T300 and c) T420, series. Figure 3.6a was taken from García-Amaya et al. [10], p. 55.

The R values (Figure 3.7) increase with the temperature of the heat treatments, and T420 series shows the highest values of 4.33, 3.69, 3.80, 4.02 and 3.91 for ZCT-0.3, ZCT-0.6, ZCT-0.9, ZCT-1.2 and ZCT-1.5, respectively. This increase in R values indicates the low symmetry around the Eu^{3+} ions and the high covalence of Eu–O, while for low R values the opposite happens [24]. The asymmetry of the matrix can be attributed to the enrichment of CdO, generated by the abundant phase crystallization of Zn, which is a highly polarizable ion that changes the crystalline field and consequently the sites where Eu^{3+} ions may be bonding.

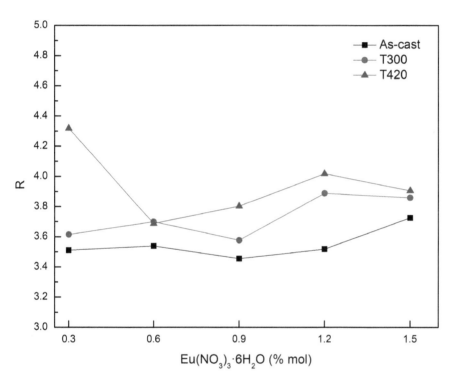

Figure 3.7: Comparison of the radius ratio (R) of the $17ZnO\bullet32CdO\bullet51TeO_2$ glasses with different Eu^{3+}-doped ZCT glasses: (a) as-cast, (b) T300 and (c) T420 series.

These values make them good candidates and show high potential for laser materials compared to the values (4.24) reported in the literature [14, 36].

As previously reported, in zinc-telluride glasses, Eu ions replace Zn sites in the vitreous reticulum, and Eu^{3+} ions become Eu^{2+} during heating [16, 39]. To investigate if this reduction phenomenon occurs, García-Amaya et al. [10] studied the luminescence spectra of T420 series glasses that were made using a He–Cd laser with an excitation wavelength of 325 nm, and the emission spectra were normalized at 405 nm (Figure 3.8[5]). The results display that the emission in this region corresponds to the intrinsic luminescence of the glass matrix, which consists of a wide band centered at ~ 450 nm and is the same for all Eu-doped glasses, and only exists as a slight increase in the intensity of the ZCT-0.15 and ZCT-0.3 glasses. These emissions are possibly related to defects in the glass matrix, or to phase crystallization. As a consequence, even after the heat treatment, the Eu^{3+} ions were not reduced.

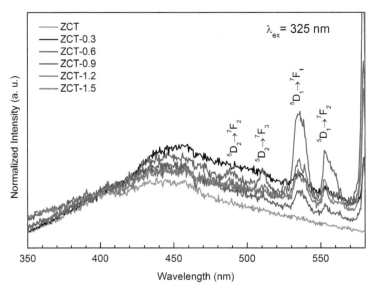

Figure 3.8: Photoluminescence spectra of the T420 series obtained at a wavelength of 325 nm. Taken from García-Amaya et al. [10], p. 55.

3.3.6 Photometric Characterization

The CIE1931 diagram (Commission International d'Eclairage) is a universal method, where all possible colors obtained by the combination of the three primary colors are represented and used to quantify the emission's wavelength hue [36]. x and y chromaticity coordinates were calculated by converting the emission spectra of glasses with different Eu^{3+} ion concentrations, obtained at $\lambda_{ex} = 292$ nm (Figure 3.9). The values of the chromaticity coordinates are shown in Table 3.6. These coordinates are used to determine the prepared glass's actual color emission, which occurs mostly in the orange-red region of the chromaticity diagram. Figure 3.9a shows that the emission color of the glasses does not vary according to the concentration of Eu^{3+}, being positioned in the same orange-reddish region except for the ZCT-0.3 glass. On the other hand, the glasses of the T300 series (Figure 3.9b) present random variations according to the concentration of Eu^{3+}. Figure 3.9c displays T420 series, which shows that with increasing Eu^{3+} ion concentration, the color coordinates shift linearly to the red region in the CIE1931 diagram. Figure 3.9a indicates that red emission predominates with increasing europium ion concentration, suggesting its suitability for red light source applications.

The correlated color temperature (CCT) was estimated with the CIE1931 color coordinates using McCarmy's formula [19, 24]:

$$\text{CCT} = -449n^3 + 3{,}525n^2 - 6{,}823n + 5{,}520.33$$

where $n = (x - x_e)/(y - y_e)$ is a line of inverted slope. Here $x_e = 0.332$ and $y_e = 0.1858$ are the epicenters. The CCT calculations for the three glass series are presented in Table 3.6.

The color purity (CP) was obtained by considering the standard coordinates of the illuminant (C) of the CIE1931 diagram ($x_i = 0.3101, y_i = 0.3162$), the emission color coordinates (x, y) and the coordinates of the dominant wavelength (x_d, y_d) concerning the coordinates (x_i, y_i):

$$\text{CP} = \frac{\sqrt{(x - x_i)^2 + (y - y_i)^2}}{\sqrt{(x_d - x_i)^2 + (y_d - y_i)^2}} \times 100\%$$

The dominant wavelength is the monochromatic wavelength of the spectrum, and the coordinates of which are on the same line that links the illuminant with the phosphorus. These values confirm that the Eu^{3+}-doped ZCT system glasses are suitable for red laser applications and display devices [24, 36], especially the T420 series.

The ZCT-1.5 glass of the T420 series is the glass with the highest potential for use as red emission phosphorus of 1,895 K, with coordinates (0.62, 0.36) and CP (96%) after excitation at 294 nm. These coordinates are close to the coordinates (0.67, 0.33) of the red phosphorus proposed by the National Television Standard Committee [5].

Table 3.6: Values of the chromatic coordinates (x, y), color temperature (CCT) and color purity (CP) obtained from glasses: (a) as-cast, (b) T300 and (c) T420 series. Photoluminescence spectrum of the 17ZnO·32CdO·51TeO₂ glasses with different Eu^{3+} doped obtained at λ_{ex} = 294 nm, a) As cast, b) T300 and c) T420, series. Figure 3.6a was taken from García-Amaya et al. [10], p. 55.

Samples	As-cast				T300				T420			
	x	y	CCT	CP	x	y	CCT	CP	x	y	CCT	CP
			K	%			K	%			K	%
ZCT-0.3	0.53	0.32	1712	59	0.53	0.34	1621	69	0.50	0.34	1669	57
ZCT-0.6	0.59	0.35	1820	85	0.56	0.35	1659	78	0.56	0.35	1638	76
ZCT-0.9	0.60	0.35	1833	87	0.57	0.36	1626	83	0.60	0.36	1789	91
ZCT-1.2	0.60	0.34	1888	85	0.61	0.36	1826	93	0.62	0.36	1865	94
ZCT-1.5	0.59	0.34	1890	84	0.62	0.36	1902	96	0.62	0.36	1895	96

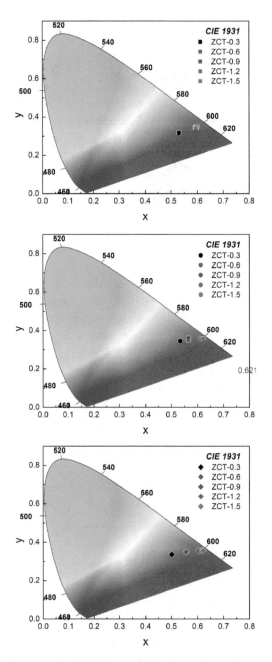

Figure 3.9: Graphic representation of the chromaticity coordinates in the diagram proposed by the CIE1931, of the 17ZnO•32CdO•51TeO$_2$ glasses with different Eu^{3+}-doped ZCT glasses: (a) as-cast, (b) T300 and (c) T420 series.

3.3.7 Decay time

García et al. [11] studied that the decay curves of the as-cast series were obtained with an emission of 615 nm after excitation at 392 nm. The decay time was speedy and is very similar for all samples. The estimated lifetime "τ" is of 0.4 ms.

3.4 Conclusion

In this chapter, we first studied green and transparent glasses. During heat treatment, coloring changed to colorless transparent at 300 °C. Finally, opaque green glass results with a thermal treatment at 420 °C. The crystalline phases of $ZnTeO_3$ and the amorphous nature of the prepared glasses were proven using XRD. The determination of the density reveals that for all the series of glasses, its behavior is nonlinear due to the changes of coordination of the groups TeO_4 to TeO_3.

The general behavior of optical absorption peaks is that they increase their intensity with a higher content of the dopant, which suggests that the higher concentration of europium increases the density of defects in the material. The optical energy band gap varies in the 2.66–3.49 eV range, and this value indicates that the glasses are semiconductors. PL confirmed that europium ions persist as Eu^{3+} after heat treatment at 420 °C, and reduction process does not occur. The vitreous matrix has intrinsic luminescence, consisting of a wide band centered at ~ 450 nm for all glasses. The intensity values of the luminescent radius (R) vary in the 3.91–4.33 range. These values make them good candidates and show a high potential for laser materials.

The ZCT-1.5 glass of the T420 series with chromaticity coordinates (0.62, 0.36) and CP of 96% has the highest potential for use as red emission phosphorus. The decay time is 0.4 ms and is very similar for all samples.

Notes

1 Table 3.1 was taken from García-Amayaet al. [10], p. 50.
2 Figure 3.2a was taken from García-Amaya et al. [10], p. 51.
3 Figure 3.2c was taken from García-Amaya et al. [10], p. 51.
4 Figure 3.6a was taken from García-Amaya et al. [10], p. 55.
5 Figure 3.8 was taken from García-Amaya et al. [10], p. 55.

References

[1] Annapoorani K., Marimuthu K. (2017). Spectroscopic properties of Eu^{3+} ions doped Barium telluroborate glasses for red laser applications. J. Non-Cryst. Solids, 463, 148–157, doi: https://doi.org/10.1016/j.jnoncrysol.2017.03.004.

[2] Annapurna Devi C. B., Mahamuda S., Swapna K., Venkateswarlu M., Srinivasa Rao A., Vijaya Prakash G. (2017). Compositional dependence of red luminescence from Eu^{3+} ions doped single and mixed alkali fluoro tungsten tellurite glasses. Opt. Mater. (Amst), 73, 260–267, doi: https://doi.org/10.1016/j.optmat.2017.08.010.

[3] Arunkumar S., Venkata Krishnaiah K., Marimuthu K. (2013). Structural and luminescence behavior of lead fluoroborate glasses containing Eu^{3+} ions. Physica. B Condens. Matter, 416, 88–100, doi: https://doi.org/10.1016/j.physb.2013.02.022.

[4] Babu S. S., Babu P., Jayasankar C. K., Sievers W., Tröster T., Wortmann G. (2007). Optical absorption and photoluminescence studies of Eu^{3+}-doped phosphate and fluorophosphate glasses. J. Lumin., 126(1), 109–120, doi: https://doi.org/10.1016/j.jlumin.2006.05.010.

[5] Caldiño U., Lira A., Meza-Rocha A. N., Camarillo I., Lozada-Morales R. (2018). Development of sodium-zinc phosphate glasses doped with Dy^{3+}, Eu^{3+} and Dy^{3+}/Eu^{3+} for yellow laser medium, reddish-orange and white phosphor applications. J. Lumin., 194(October 2017), 231–239, doi: https://doi.org/10.1016/j.jlumin.2017.10.028.

[6] De Almeida R., Da Silva D. M., Kassab L. R. P., De Araújo C. B. (2008). Eu^{3+} luminescence in tellurite glasses with gold nanostructures. Opt. Commun., 281(1), 108–112, doi: https://doi.org/10.1016/j.optcom.2007.08.072.

[7] Dehelean A., Rada S., Kacso I., Culea E. (2013). IR, UV-vis spectroscopic and DSC investigations of europium doped tellurite glasses obtained by sol-gel synthesis. J. Phys. Chem. Solids, 74(9), 1235–1239, doi: https://doi.org/10.1016/j.jpcs.2013.03.022.

[8] El Mallawany R. (2012). Tellurite Glasses Handbook: Physical Properties and Data. Second Edi, Taylor & Francis Group, Ed., Kingdom of Saudi Arabia Boca.

[9] Eraiah B. (2006). Optical properties of samarium doped zinc – Tellurite glasses. Bull. Mater. Sci., 29(4), 375–378, doi: https://doi.org/10.1007/BF02704138.

[10] García-Amaya IV., Zayas M. E., Alvarado-Rivera J., Cortez-Valadez M., Pérez-Tello M., Cayetano-Castro N., Mendoza-Córdova A. (2018). Influence of Eu_2O_3 on phase crystallization and nanocrystals formation in tellurite glasses. J. Non-Crystalline Solids, 499(January), 49–57, doi: https://doi.org/10.1016/j.jnoncrysol.2018.07.018.

[11] García IV., Zayas M. E., Alvarado J., Álvarez E., Gallardo-Heredia S. A., Limón G. A., Rincón J. M. (2015). Spectroscopic studies of the behavior of Eu^{3+} on the luminescence of cadmium tellurite glasses. J. Spectroscopy, 2015, 1–8, doi: https://doi.org/10.1155/2015/478329.

[12] Ghribi N., Dutreilh-colas M., Duclère J., Hayakawa T., Carreaud J., Karray R., Kabadou A. (2015). Thermal, optical and structural properties of glasses within the TeO_2–TiO_2–ZnO system. J. Alloys. Compd., 622, 333–340, doi: https://doi.org/10.1016/j.jallcom.2014.10.063

[13] Hirashima H., Ide M., Yoshida T. (1986). Memory switching of $V_2O_5TeO_2$ glasses. J. Non-Cryst. Solids, 86(3), 327–335, doi: https://doi.org/10.1016/0022-3093(86)90021-9.

[14] Karunakaran R. T., Marimuthu K., Surendra Babu S., Arumugam S. (2009). Structural, optical and thermal studies of Eu^{3+} ions in lithium fluoroborate glasses. Solid State Sci., 11(11), 1882–1889, doi: https://doi.org/10.1016/j.solidstatesciences.2009.08.001.

[15] Kumar A., Rai D., Rai S. (2002). Optical studies of Eu^{3+} ions doped in tellurite glass. Spectrochimica Acta Part A, 58(10), 2115–2125, doi: https://doi.org/10.1016/S1386-1425(01)00684-9.

[16] Lian Z., Wang J., Lv Y., Wang S., Su Q. (2007). The reduction of Eu^{3+} to Eu^{2+} in air and luminescence properties of Eu^{2+} activated ZnO-B_2O_3-P_2O_5 glasses. J. Alloys. Compd., 430(1–2), 257–261, doi: https://doi.org/10.1016/j.jallcom.2006.05.002.

[17] Maheshvaran K., Marimuthu K. (2012). Concentration dependent Eu^{3+} doped boro-tellurite glasses-Structural and optical investigations. J. Lumin., 132(9), 2259–2267, doi: https://doi.org/10.1016/j.jlumin.2012.04.022.

[18] Marzouk M., ElBatal H., Eisa W. (2016). Optical stability of 3d transition metal ions doped-cadmium borate glasses towards γ-rays interaction. Indian J. Phys., 90(7), 781–791, doi: https://doi.org/10.1007/s12648-015-0804-7.

[19] McCamy C. S. (1992). Correlated color temperature as an explicit function of chromaticity coordinates. Color Res. Appl., 17(2), 142–144, doi: https://doi.org/10.1002/col.5080170211.

[20] Walas M., Lewandowski T., Anna Synak M. Ł., Wojciech Sadowski B. K. (2017). Eu^{3+} doped tellurite glass ceramics containing SrF_2 nanocrystals: Preparation, structure and luminescence properties. J. Alloys. Compd., 696, 619–626, doi: https://doi.org/10.1016/j.jallcom.2016.11.301

[21] Mohamed E. A., Ahmad F., Aly K. A. (2012). Effect of lithium addition on thermal and optical properties of zinc–tellurite glass. J. Alloys. Compd., 538, 230–236, doi: https://doi.org/10.1016/j.jallcom.2012.05.044

[22] Mohan S., Singh D. P., Kaur S. (2015). Structural and optical investigations of CdO-Na_2CO_3-H_3BO_3 glasses. 801(July 2014), 796–801, doi: https://doi.org/dx.doi.org/10.1139/cjp-2014-0404.

[23] Mohan S., Thind K .S. (2017). Optical and spectroscopic properties of neodymium doped cadmium-sodium borate glasses. Opt. Laser Technol., 95, 36–41. doi: https://doi.org/10.1016/j.optlastec.2017.04.016.

[24] Pravinraj S., Vijayakumar M., Marimuthu K. (2017). Enhanced luminescence behaviour of Eu^{3+} doped heavy metal oxide telluroborate glasses for Laser and LED applications. Physica. B Condens. Matter, 509(January), 84–93, doi: https://doi.org/10.1016/j.physb.2017.01.008.

[25] Pulluru C. R., Kalluru R. R., Reddy B. R., Konovalova T. A., Kispert L. D. (2005a). Persistent spectral hole burning in europium-doped sodium tellurite glass. Appl. Phys. Lett., 87, 1–3. doi: https://doi.org/10.1063/1.2035884.

[26] Pulluru C. R., Kalluru R. R., Reddy B. R., Konovalova T. A., Kispert L. D. (2005b). Persistent spectral hole burning in europium-doped sodium tellurite glass. Appl. Phys. Lett., 87(9), 1–4, doi: https://doi.org/10.1063/1.2035884.

[27] Rodríguez-Carvajal D. A., Meza-Rocha A. N., Caldiño U., Lozada-Morales R., Álvarez E., Zayas M. E. (2016). Reddish-orange, neutral and warm white emissions in Eu3+, Dy3+ and Dy3+/Eu3+ doped CdO-$GeO2$-$TeO2$ glasses. Solid State Sci., 61, 70–76, doi: https://doi.org/10.1016/j.solidstatesciences.2016.09.009.

[28] Rolli R., Montagna M., Chaussedent S., Monteil A., Tikhomirov V. K., Ferrari M. (2003). Erbium-doped tellurite glasses with high quantum efficiency and broadband stimulated emission cross section at 1.5 μm. Opt. Mater. (Amst), 21(4), 743–748, doi: https://doi.org/10.1016/S0925-3467(02)00092-7.

[29] Ruvalcaba-Cornejo C., Flores-Acosta M., Zayas M., Lozada-Morales R., Palomino-Merino R., Espinosa J. E., Soto A. B. (2008). Photoluminescence properties of the ZnO-CdO-TeO_2 system doped with the Tb^{3+} and Yb^{3+} ions. J. Lumin., 128(2), 213–216, doi: https://doi.org/10.1016/j.jlumin.2007.07.004.

[30] Ruvalcaba-Cornejo C., Lozada-Morales R., Zayas M. E., Rincón J. M., Marquez H., Flores A. (2013). Effect of the Eu^{3+} Addition on photoluminescence and microstructure of ZnO–CdO–TeO_2 Glasses. J. Am. Ceram. Soc., 96(10), 3084–3088, doi: https://doi.org/10.1111/jace.12542.

[31] Ruvalcaba-Cornejo C., Zayas M. E., Castillo S. J., Lozada-Morales R., Pérez-Tello M., Díaz C. G., Rincón J. M. (2011). Optical and thermal analysis of ZnO–CdO–TeO$_2$ glasses doped with Nd^{3+}. Opt. Mater. (Amst), 33(6), 823–826, doi: https://doi.org/10.1016/j.optmat.2011.01.001.

[32] Sołtys M., Janek J., Zur L., Pisarska J., Pisarski W. A. (2015). Compositional-dependent europium-doped lead phosphate glasses and their spectroscopic properties. Opt. Mater. (Amst), 40, 91–96. doi: https://doi.org/10.1016/j.optmat.2014.11.052.

[33] Stambouli W., Elhouichet H., Gelloz B., Férid M. (2013). Optical and spectroscopic properties of Eu-doped tellurite glasses and glass ceramics. J. Lumin., 138, 201–208, doi: https://doi.org/10.1016/j.jlumin.2013.01.019.

[34] Stavrou E., Tsiantos C., Tsopouridou R. D., Kripotou S., Kontos A. G., Raptis C., Khatir S. (2010). Raman scattering boson peak and differential scanning calorimetry studies of the glass transition in tellurium-zinc oxide glasses. J. Phys. Condens. Mat., 22(19), doi: https://doi.org/10.1088/0953-8984/22/19/195103.

[35] Van Deun R., Binnemans K., Görller-Walrand C., Adam J. L. (1998). Optical properties of Eu^{3+}-doped fluorophosphate glasses. J. Phys. Condens. Mat., 10(98)7231–7241. Retrieved from https://lirias.kuleuven.be/bitstream/123456789/33475/1/33Euglass.pdf.

[36] Vijayakumar R., Maheshvaran K., Sudarsan V., Marimuthu K. (2014). Concentration dependent luminescence studies on Eu^{3+} doped telluro fluoroborate glasses. J. Lumin., 154, 160–167, doi: https://doi.org/10.1016/j.jlumin.2014.04.022.

[37] Wang F., Shen L. F., Chen B. J., Pun E. Y. B., Lin H. (2013). Broadband fluorescence emission of Eu^{3+} doped germanotellurite glasses for fiber-based irradiation light sources. Opt. Mater. Express, 3(11), 1931, doi: https://doi.org/10.1364/OME.3.001931.

[38] Wang R., Zhou D., Qiu J., Yang Y., Wang C. (2015). Color-tunable luminescence in Eu^{3+}/Tb^{3+} co-doped oxyfluoride glass and transparent glass-ceramics. J. Alloys. Compd., 629, 310–314, doi: https://doi.org/10.1016/j.jallcom.2014.12.233

[39] Xia H., Nie Q., Zhang J., Wang J. (2003). Preparation and optical spectroscopy of Na$_2$O-TeO$_2$-ZnO glasses containing divalent europium ions. Mater. Lett., 57(24–25), 3895–3898, doi: https://doi.org/10.1016/S0167-577X(03)00236-2.

[40] Yu K. M., Detert D. M., Chen G., Zhu W., Liu C., Grankowska S., Walukiewicz W. (2016). Defects and properties of cadmium oxide based transparent conductors. J. Appl. Phys., 119(18), 1–10, doi: https://doi.org/10.1063/1.4948236.

[41] Zayas M. E., Arizpe H., Castillo S. J., Medrano F., Diaz G. C., Rincón J. M., Espinoza F. (2005). Glass formation area and structure of glassy materials obtained from the ZnO–CdO–TeO$_2$ termary system. Phys. Chem. Glasses, 46(1), 46–50.

[42] Zhou L. Y., Wei J. S., Yi L. H., Gong F. Z., Huang J. L., Wang W. (2009). A promising red phosphor MgMoO$_4$: Eu^{3+} for white light emitting diodes. Mater. Res. Bull., 44(6), 1411–1414, doi: https://doi.org/10.1016/j.materresbull.2008.11.019.

S. P. Kori, R. Rohatgi, V. Lahariya

Chapter 4
Nanophosphors: Emerging Materials for Forensic Applications

Abstract: The amalgamation of nanotechnology with latent fingermark detection increases its utility in forensic science. Nanoparticles are frequently used to develop latent fingermarks in a wide range of porous, semiporous and nonporous surfaces due to their high sensitivity and selectivity toward the different components of residues. The utility of these particles can be enhanced by doping them with fluorescent dyes. Cadmium sulfide-, zinc oxide-, zinc sulfide-, titanium dioxide- and aluminum oxide-based nanoparticles are widely used to develop weak and fragmented chance prints of forensic importance. In addition to this, rare earth oxides such as europium oxide- and cesium oxide-based composition are used to enhance the visibility of latent marks. These compositions can also be used to develop latent fingermarks on moist and sticky surfaces of adhesive tapes. The use of these compositions not only detects the latent fingermarks but also improves the contrast of developed prints. The application of these fluorescent materials is not only limited to fingermarks detection. These fluorescent nanoparticles can also be used to detect the presence of body fluids in crime scene or items recovered from the crime scene. In addition to this, these fluorescent compounds are also used in certain kinds of printing and writing inks and for the identification purpose. These compounds are used as tag compounds in ink.

This chapter provides a comprehensive review of the recent research related to different kinds of fluorescent materials, including their different novel characteristics and potential applications in forensic science as fingerprint detection. We begin with a historical background of luminescence, its mechanism with different excitation and parameters, working principle and different fluorescent nanomaterials used for developing and detecting latent fingerprint investigations. The recent progress on the improvement of their characteristics and finding novel potential applications such as the latent fingerprint detection is discussed.

Keywords: luminescence, semiconductor nanomaterials, fluorescent nanomaterials, forensic applications, latent fingerprints

4.1 Introduction

With the advancement of science and technology, nanotechnology has been used as a tool in the field of forensic science. It has brought many novel ideas for developing and detecting fingerprints. Application of nanomaterials in forensic science aims at identifying, individualizing and evaluating crime scene evidence. Fingerprint is measured as an important evidence for forensic investigations, and nano-based materials put enormous future potential in this field. Fingerprints obtained from crime scene aid in identifying the suspect and bring them to justice. Nanoforensics or nanomaterials used in aiding criminal investigations have become indispensable. For instance, the use of nanomaterials in fingerprint detection, markers in DNA profiling, biological fluid detection, drugs and arson, explosives and firearm examinations have proved their utility, relevance and advantage over other techniques tremendously. Some of the noteworthy applications of nanomaterials in forensic science, in general, have been shown in Figure 4.1. There has been a vast research undertaken in the development of latent fingerprints in combination with fluorescent dyes and nanomaterials and nanophosphors. Recently, nanophosphors are considered a good choice in forensic application, and to realize the improved performance of nanophosphors over bulk material, however, detailed investigation is essential. Fluorescent nanomaterials such as quantum dots (QDs) and upconversion nanomaterials have attracted immense interest and promising results in fingerprint detection [26]. Moreover, these kind of materials exhibit high intensity, photo and chemical stability, ease of operation and reliability.

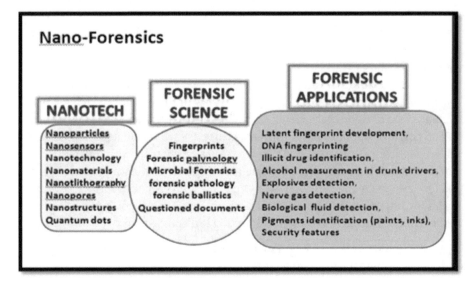

Figure 4.1: Role of nanomaterials in forensic science.

Materials having a particle size in nanoregime excited by high-energy photon emit radiation in visible region due to the electronic transitions between the levels of impurity ions inside the band gap of the host are called nanophosphors. Basically, phosphor term is used to describe luminescent materials. Nowadays, nanophosphor is one of the most promising and reliable emerging materials to investigate and visualize the fingerprint evidence in forensic science applications. It not only improves the quality of the developed prints but also enhances the visual ridge characteristics remarkably. In particular, semiconductor nanophosphor materials have wide applications in the field of forensic science in the sense of developing and detection of latent fingerprints.

Keywords: luminescence, latent fingerprint, phosphor

Therefore, this chapter focuses on applications of semiconductor nanoparticles, rare-earth-doped nanophosphor and other nanophosphors as powder or liquid reagent in forensic science. A latent print is an impression left by the palmer side of the friction skin phalanges. The identification and lifting of chance prints from crime scene are an integral part of criminal individualization. Perspiration from sweat glands causes the formation of fingerprints on various surfaces. Groomed fingerprints can be obtained by using roller-slab or ink-pad method for taking fingerprints on ten-digit fingerprint record cards. Care must be taken to record the fingerprint from the proximal end to the distal end. Forensic scientists have used fingerprints in crime scene inspections. A person's fingerprints do not change over time. The friction ridges that create fingerprints are formed in the first trimester of the gestation. Fingerprints can be used in providing biometric security and in identifying amnesia victims and unknown deceased [18, 29, 65].

4.1.1 Types of Fingerprints

The infinitesimal details of fingerprints like bifurcations, islands and enclosures are known as Galton details. The Galton details have a significant role for individualization since it varies from person to person. It can be classified mainly into three types (Figure 4.2) [18].
(i) Latent fingerprints
(ii) Patent fingerprints
(iii) Plastic fingerprints

(i) *Latent fingerprints:* Such fingerprints are invisible for naked eyes. However, it may be observed by powder dusting method. It can leave on a surface by deposits of oils and/or perspiration from the finger [18].
(ii) *Patent fingerprints:* It is a visible impression of individual's fingertip left on a surface as a result of residue on the finger. It can also be developed by the blood oil or ink and on the fluid like paint and grease. These are mostly photographed in oblique lighting for visualization and enhancement [18].
(iii) *Plastic fingerprints:* It is relatively easier for visualization from naked eyes. This three-dimensional impression can be obtained by pressing your fingers in

paint, wax, soap or tar. Additional physical or chemical method for development is not required [18].

Keywords: latent fingerprint, patent fingerprint, plastic fingerprint

Figure 4.2: Type of fingerprints. Reprint permission from Edgar [18].

4.1.2 Patterns of Fingerprints

There are three basic patterns of fingerprint ridges as shown in Figure 4.3.
i. **Arch:** The ridges enter from one side of the finger, rise in the center forming an arc and then exit on the other side of the finger. It appears as a wave-like pattern and comprises plain arches and tented arches. Tented arches rise to a sharper point than plain arches. It builds up to about 5% of all pattern types [65].
ii. **Loop:** The edges enter from one side of a finger, form a curve and then exit on that same side. In addition, prints recurve back on themselves to make a loop pattern. It is further divided into radial loops (pointing toward the thumb or radius bone) and ulnar loops (pointing toward the ulna bone), approximately 60% of pattern type belong to this [18, 29, 65].
iii. **Whorl:** Usually whorls can have four types: plain, central pocket loop, double loop and accidental loop. Plain whorls are concentric circles, central pocket loops are represented by a loop with a whorl at the end while accidental loops are irregular in shape and double loops form an S-like pattern. Whorls make up about 35% of pattern types (Edgar et al., 2012). Such circularly ridges form around a central point on the finger or spiral patterns like tiny whirlpools.

The conventionally used techniques in detecting fingerprints such as powder methods, ninhydrin, cyanoacrylate glue fuming, silver nitrate method, small particle reagent (SPR) method and phase transfer catalyst have all been admitted with some drawbacks and long-term use of hazards by the experts. Some of these drawbacks include necessity of sunlight for reaction completion as in case of silver nitrate, while some require the use of ALS (alternate light sources) for print enhancement. However, the general downside of all these conventional methods of developing fingerprints is high toxicity, low contrast, selectivity and sensitivity.

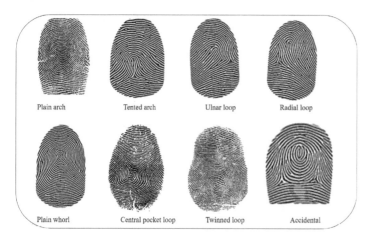

Figure 4.3: Patterns of fingerprints. Reprint permission from [29].

To overcome this disadvantage of conventional methods of fingerprints development, many efforts have been paid to the use of fluorescent nanomaterials, semiconductor QDs and rare earth upconversion fluorescent nanomaterials due to their remarkable optical and chemical properties [70]. The development of latent fingerprints using fluorescent nanomaterials can achieve higher contrast, distinct and clear ridge details and lesser toxicity in comparison to previously used methods.

4.2 Luminescence

Luminescence means spontaneous emitted photon in the visible region. It was first discovered by a German physicist in the eighteenth century. Depending on the means of excitation, luminescence is classified into different forms. Luminescence is an interdisciplinary subject, which is associated with the applied science. In general, luminescence is the phenomena of spontaneously visible light emission by a substance. It is known as cool light source, which means light emitted by sources except a hot, incandescent body such as firefly, oscilloscope screen, fluorescent tube and television screen. It is caused by applying an external energy source to a substance, which undergoes an electronic transition from higher energy levels to the ground state.

In case of luminescence, the prefix indicates the external energy source that is used to excite the luminescence called energy conversion-type luminescence such as photoluminescence, electroluminescence and bioluminescence are excited by light, electric field and biological organism, respectively. On the other hand, in some luminescence, prefix shows the process to stimulate the luminescence; for example, thermoluminescence, chemiluminescence and sonoluminescence. Thermoluminescence occurs by absorbing ionizing radiation, while crystalloluminescence can be observed through crystallization. Mechanoluminescence is the emission of light by inducing mechanical pressure [7]. The process may occur in crystalline and noncrystalline solids as well as liquids and gases. Furthermore, depending on the delay time between the excitation and emission, luminescence can be classified into two categories. If the storage time is shorter than 10^{-8} s, the process is termed as fluorescence; otherwise, it is phosphorescence [7]. The phosphorescence also known as delayed luminescence. Materials with high radiative recombination probability are usually referred to as phosphors [7].

The interval 10^{-8} s is selected as the lifetime of an atom in the excited state, which comes back to the ground state accompanied by dipole relaxation [7, 31]. Photoluminescence is a process in which a substance absorbs a photon in the visible region, exciting electrons to a higher electronic excited state, and then radiates a photon as the electron returns to a lower energy state. If the molecule undergoes internal energy redistribution after the initial photon absorption, the radiated photon is of longer wavelength (i.e., lower energy) than the absorbed photon called Stoke's shift [7]. Fluorescence and phosphorescence are special kinds of photoluminescence. The configuration coordinate model is used to explain the mechanism of luminescence as shown in Figure 4.4, where a coordinate represents the potential energy of ground state and excited state. The parabolic curve represents the atomic group that behaves like a simple harmonic oscillator. By absorbing the energy, electron gets excited and moves vertically to the excited state. After the absorption, nuclei do not stay in the equilibrium position and move to the minimum position of excited state. This process is known as vibrational relaxation. Therefore, lattice vibration or phonon has been dissipated corresponding to this transition. After reaching the minimum position of excited state, the electron comes to the ground state emitting the radiation.

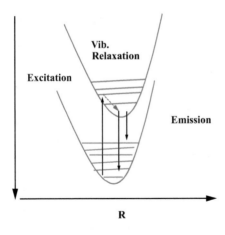

Figure 4.4: Configurational coordinate diagram of photoluminescence.

4.3 Semiconductor Nanomaterials

The special functionality of materials that remarkably enhances in the nanosize regime (10^{-9} m) is known as nanomaterials. In nanomaterials, the short dimension length must be smaller than the phase coherent length or the mean free path scattering length for electrons in the materials. When the reduced dimension of materials is comparable to the exciton Bohr radius, the quantum size effect has been observed. In this case, the electron and holes movement restrict for definite values of the energies. Thus, the continuum of electronic energy states in conduction and valence bands is split into discrete energy levels with energy spacing relative to the band edge. In consequence, highest occupied valence band and lowest unoccupied conduction band are shifted to more negative and positive values, respectively, resulting in widening of band gap.

Keywords: semiconductor, nanomaterials, fluorescence

The ability to tune the absorption band edge and consequently the fluorescent emission wavelength via specific nanocrystal size which control the confinement energy of the charge carriers is one of the significant properties of nanocrystals [55]. The size-dependent properties seem to have reduced particle size comparable to the exciton Bohr radius in the bulk material. With decreasing particle size, the energy band gap and absorption edge shifts toward higher energy side accompanied by the appearance of a strong excitonic peak. Due to its small size and relatively large surface-to-volume ratio, the effect on the charge carriers cannot be neglected [2]. The optoelectronic as well as luminescent properties can be transformed in nanoscale. The nanometer-sized particles explore many properties such as size quantization, surface effect, lattice contraction, unusual fluorescence and enhanced oscillator strength, which are smaller than the diameter of their corresponding bulk excitons strongly depend on the size of the particles due to quantum confinement of the excited states [19]. By precise monitoring of the size and surface effect, the desired optical, magnetic, elastic and chemical properties from the material may be possible. Moreover, the surface of semiconductor nanoparticles is more sensitive and defective. Therefore, it is keenly interesting to study the properties of nanomaterials. The energy band diagram is shown in Figure 4.5. This band structure depicts the electronic band structure in bulk, nanoscale particle and at molecular level. In bulk structures, electronic and optical properties can only be altered by adding constituent's impurity. It can create some defects, interstitials or substitutions in the material.

In semiconductor nanomaterials, optical and electronic properties can be precisely tuned by changing the size of the particle in addition to adding dopants. Due to its unique properties, semiconductor nanomaterials have some promising and interesting potential applications such as biological, labeling, sensor, photocatalysis, phosphor and fingerprint detection [3, 6, 35]. Moreover, the advantage of using such nanomaterials is opposed to organic dyes that are currently used, and QDs are brighter and more resistant to photobleaching [62].

Keywords: band structure, luminescence, nanoparticles, forensic applications

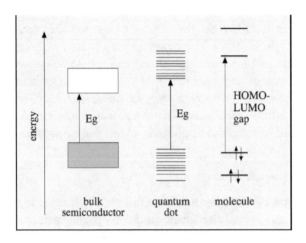

Figure 4.5: Energy band structure.

QDs are such nanoparticles whose size is comparable to the exciton Bohr radius. The exciton Bohr radius depends on the effective mass of electron and hole and dielectric constant, which is different for different materials. When the size of QD is below the exciton Bohr radius, it is referred to as strong confinement, and if the size is greater than the exciton Bohr radius it is said to have weak confinement effect. Due to its exceptional properties, these lead to some promising potential applications. The QDs are more photostable under UV excitation as comparable to organic dye and also less toxic. The absorption and emission can be tuned with reducing size and stable fluorescence.

4.3.1 Luminescence in Semiconductor Nanoparticles

Luminescence properties of nanocrystalline semiconductor exhibit band edge luminescence or excitonic recombination with smaller crystalline size. The intensity of redshifted band has been increased with a decrease in crystalline size. In nanocrystal, atoms are present on the surface where dangling bonds create trap states. These traps influence the electron/hole recombination parameters that affect the luminescence. Furthermore, luminescence can also be modified by the doping of impurity ions and ion attached to the nanocrystal surface. In such phenomena, addition of impurities disturbs the crystal structure and symmetry, thereby additional localized levels present in the forbidden gap. In solids, such sites are known as luminescence centers. Such centers play an important role in order to enhance the luminescence. If the material irradiates by low energy photon, the intensity of emission subsequently increases. This process is known optically stimulation. Conversely, the process is known as quenching. However, quenching can be occurred by concentration effect or temperature effect. These are called concentration quenching and temperature quenching, respectively.

4.3.2 Forensic Applications of Semiconductor Nanoparticles

Fingerprints have been utilized as an exceptional confirmation, which were later additionally utilized on Babylonian mud tablets for business exchange. With time, many different techniques were employed to develop fingerprints using powders and chemicals. Fingerprint powders usually adhere to the sweat residue left by the fingertips and show distinct fingerprint patterns that help in personal identification.

The body secretions like sweat, together with surrounding contaminations secreted from any of the three types of glands vis-à-vis apocrine, eccrine and sebaceous glands, constitute to form latent prints [9]. The most encountered are the latent fingerprints at the scene of crime. It is invisible in nature, and therefore, the most challenging is to identify and develop [77].

The selection of technique for latent fingerprint development rests on the nature of the surface and weather conditions. Latent fingerprints are commonly developed by various colored materials. However, its adherence power is the main problem, which is not able to provide a clear image of the fingerprint, so identification is a major issue. The decision of a method for latent unique finger impression advancement relies upon the substance of the dormant finger impression just as on the idea of the surface. The disadvantage of these materials is its adherence power that is hard to get away from the unique finger impression, so distinguishing proof is a significant issue. In the mid-1970s, some data were accessible on the recognition of latent fingerprints with nanoparticles.

Ag and Au nanoparticles were utilized in the normal silver physical developer methods for the representation of inactive fingerprints on permeable paper for more than two decades. In any case, during this time, they were not used much for the recognition of inert fingerprints. From the year 2000, semiconductor nanoparticles and uncommon earth-doped nanoparticles became well known for the discovery of dormant fingerprints on permeable and nonporous surfaces. The utilization of nanoparticles has as of late demonstrated an extraordinary potential in delivering the up-and-coming age of unique mark improvement methods known as nanofingerprints. One of the most significant confirmations found at the wrongdoing scene is fingerprints since they are perceived as summed-up verifications of a human character [40]. The confinements on the identification of inactive unique finger impression found in wrongdoing scenes can be improved by the great emanation of semiconductor nanoparticles.

Additionally, recognition capacity on inactive fingerprints is required to improve because the semiconductor nanomaterials can collaborate with the outside of unique finger impression deposits, in this manner prompting better affectability and selectivity in inert unique finger impression location. There could be many distinctive nanoparticles that can be applied on the latent fingerprints for enhanced visualization using oxides of Zn, Si, Ag, Al and Eu, fluorescent C and amphiphilic silica on the scope of particle surfaces that are most usually utilized. These nanoparticles help in bringing out the edge clarity and sharpness in developed prints [20].

Moreover, QDs are perceived as a colorific semiconductor nanomaterial as they can radiate light under the UV or blue excitation. It exhibits wide adsorption spectra, size-subordinate tunable discharge, thin outflow phantom, transfer speed and high quantum yield. In this connection, Menzel and colleagues revealed a model that utilized QDs for latent finger impression [46, 47]. Therefore, QDs emerged new opportunities for development and detection of latent fingerprint with best contrast, sensitivity and selectivity. Previously, cadmium sulfide (CdS), CdSe and CdTe QDs have been used for developing latent fingerprints [Menzel et al. 2000, 70]. Mostly, QDs are used in the powder dusting method, where they get adhered to the aqueous or oily composites of sweat sebum. This adherence largely resembles the manner of the typical powder dusting method. The physical adsorption and electrostatic attraction are the two most important factors on which the interaction between QDs and latent prints depends greatly. The emitted fluorescence from the QDs is significantly low, which causes a poor contrast to high oxidation behavior of QDs that are exposed in atmosphere as a result of adherence to latent prints. This indicates that the long-lasting protection of unique finger impression is another issue. To overcome this problem, QD-based nanocomposites were often used [70]. Moreover, QDs that develop latent fingerprints in liquid medium were made conceivable by the property to selectively stick to certain portions of sweat residues that are under liquid medium at crime scene [70]. QDs are modified by various specific ligands with functional groups to achieve the adsorption interaction [70]. The specific assimilation collaboration among QDs and inert unique finger impression buildups is controlled by compound responses. Notwithstanding some physical procedures, for example, physical adsorption and electrostatic fascination may happen simultaneously.

To achieve the assimilation cooperation, QDs were regularly altered by different explicit ligands with useful gatherings. The different kinds of nanoparticles and their applications in the identification of latent fingerprints are discussed in this section. Most widely used materials for fingerprint detection in forensic science are CdS-, zinc sulfide (ZnS)-, silicon dioxide (SiO_2)-, titanium dioxide (TiO_2)-, zinc oxide (ZnO)- and aluminum-based nanoparticles.

4.3.2.1 Cadmium Sulfide (CdS) Nanoparticles

Direct band gap semiconductors are ideal materials for optical and luminescence studies. In case of CdS, visible luminescence can only originate from transition involving localized states. CdS QDs are most widely employed semiconductors due to their chemical, physical and optoelectronic properties which make them potentially useful in forensic, catalytic and bioanalytical applications [24]. They have a band gap of 2.42 eV (512 nm) in the visible region. Absorption in bulk CdS is excitonic in nature having binding energy of exciton ~28 meV [58]. The Bohr exciton radius is 5.8 nm; therefore, strong quantum size effect appears with the crystallite size around 5–6 nm and less [74]. Sooklal et al. firstly prepared CdS QDs using polyamidoamine dendrimers as a host material [63]. This method is advantageous over other methods in that they require lesser synthetic conditions for reaction and the various reactions on the surface of dendrimers can be utilized to provide the complex that are more soluble in any solvent. CdS/dendrimer nanocomposites may be useful for fabricating novel optical sensors and microelectronic devices, and in forensic science [30, 63] and luminescent biological probes [32]. The incentive for using CdS/dendrimer in fingerprints identification is that these are robust but flexible in attaching to conjugate functional groups for selective tagging of fingerprints by both physical and chemical methods [30].

In 2000, the first study on CdS nanocrystals modified with dioctyl sodium sulfosuccinate (DSS) in heptane for developing latent fingerprints on metallic and sticky substrates was reported [46, 47]. Firstly, latent prints were pre-fumed with cyanoacrylate ester followed by immersion in CdS/DSS nanocrystals in organic solution for a few minutes.

Keywords: CdS nanoparticles, development, fingerprint

The resultant samples were gently rinsed with hexane and air dried. The developed prints deposited on aluminum foil and soft drink cans showed high contrast while unfumed bare fingerprints deposited on metal, glass and plastic surfaces were rendered unsuccessful. The fact is that nonpolar organic solvents can obliterate oily residue present in fingerprints [46, 47].

Later study reported that CdS/chitosan nanocrystals can be useful for developing the latent fingerprints on aluminum substrates [17]. They transformed the synthesized nanocrystals into powder form and subsequently latent fingerprints were fumed with cyanoacrylate ester. In this process, the developed pattern exhibited an improved contrast, although sensitivity and selectivity were poor (as shown in Figure 4.6). It was due to certainly aggregation of CdS/chitosan nanocrystals by the freeze-drying process. Hence, the quality of latent fingerprint was not much improved.

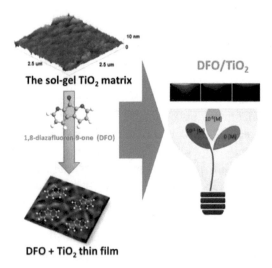

Figure 4.6: Latent print pictures on aluminum substrate under 450 nm excitation by (a) 550 nm and (b) 565 nm filter (CdS/chitosan NCs). Reproduced with permission [17]. Copyright 2009, Elsevier.

Thereafter, Algarra and coworkers observed latent fingerprints deposited on various types of substrates by using CdS/PPH nanocrystals [1]. The chemically functionalized phosphate with mercaptopropyl hybrid heterostructures was used to synthesize the CdS/PPH nanocrystals. Afterward, the as-synthesized CdS/PPH nanocrystals were applied for development of the latent fingerprints on plastic, glass, steel, ceramic and wood surfaces. It was observed that the developed latent prints have good contrast under the excitation of 450 nm light. Though the sensitivity and selectivity were quite low, efforts made on this aim was ineffective due to the background interference effect.

4.3.2.2 Zinc Sulfide Nanoparticles

ZnS is a compound semiconductor material. Due to its low absorption in the visible range and high refractive index (about 2.2), it is widely used as efficient phosphor materials for various applications [23]. ZnS doped with transition metal or rare earth exhibits attractive fluorescent properties with increased optically active sites for optical as well as forensic applications.

Keywords: ZnS, CdS, nanocomposite, latent fingerprints

It has a band gap of 3.68 eV at 300 K with excitation on the Bohr diameter 5.2 nm which corresponds to UV radiation for optical interband transition with a wavelength of 340 nm [59]. Basically, it occurs in two types: cubic (sphalerite or zinc blende) structure and hexagonal (wurtzite) structure. The sphalerite structure can be found in temperature resulting from a cubic close packing of ions, while the wurtzite structure can be observed in high temperature, derived from a hexagonal close packing scheme [75].

Petroleum ether solution of CdSe/ZnS nanoparticles stabilized by octadecaneamine with slight modification. Si and paper-thin layer behavior impressed sebaceous fingermarks were absorbed in a petroleum ether solution of CdSe/ZnS and, after air dried samples observed under ultraviolet (UV) irradiation. Well-developed images could be found on the Si thin layer. Optical microscopy study confirmed the aggregation of CdSe/ZnS nanoparticles on the fingermark ridges [34]. Hence, it provides an alternate method for envisaging latent fingermark on wet nonporous substrate. In addition, no prerequisite for posttreatment was required. Further, 3-mercaptopropionic acid–modified Cu-doped ZnS nanoparticles in aqueous solution was used for development of blood fingerprints deposited on various nonporous substrates, including transparent polypropylene, black polypropylene, glass and aluminum foil [49]. Cu-doped ZnS nanoparticles have been used for the development of blood fingerprints due to a sufficiently high contrast, sensitivity and selectivity. Becue and coworkers found the similar results and explained the mechanisms and procedures for it [4]. Hence, the study on the development of blood fingerprints using Cu-doped ZnS nanoparticles was more potent than AY 7 and CdTe nanoparticles (as shown in Figure 4.7), although the fingerprints developed by Cu-doped ZnS nanoparticles on the transparent polypropylene materials were not clearly visible due to the background fluorescent interference. In this regard, Xu et al. have developed imaging of eccrine latent prints with nontoxic ZnS:Mn QDs. ZnS:Mn QDs with different surface-topping ligands and in high temperature, show natural orange fluorescence discharge. ZnS:Mn quantum dots plainly exposed the optical pictures of eccrine latent prints. Moreover, N-acetyl-cysteine-topped ZnS:Mn QDs were able to recolor the eccrine latent fingerprints within 5 s swiftly. In the interim, the level 2 and 3 foundations of fingerprints could be similar and simply distinguished.

However, without QDs or without scouring and stepping the finger onto foil, no fluorescent unique mark pictures could be imagined [78]. Besides, new unique mark, matured (5, 10 and 50 days), inadequate eccrine latent prints could also be effectively recolored with N-acetyl-cysteine-topped ZnS:Mn QDs, representing the diagnostic capability of this technique in forensic applications. The technique was very useful for the enhancement of latent fingerprint on progression of nonporous surfaces such as aluminum foil, glass and dark plastic packs.

Even, it was able to visualize the matured (as long as 50 days) and inadequate latent prints effectively by recoloring with N-L-Cys-topped ZnS:Mn QDs. This technique is promising for possible applications at the scene of a crime for forensic investigations [78].

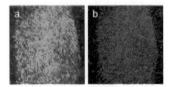

Figure 4.7: Development of blood fingerprint by using Cu-doped ZnS nanoparticles: AY7, and CdTe quantum dots under 300–400 nm excitation on various surfaces of glass, transparent polypropylene lack polyethylene and aluminum a, a′, e, e′: b, b′, f, f′: c, c′, g, g′: d, d′, h, h′, respectively. Reproduced with permission from [4]. Copyright 2013, Elsevier.

4.3.2.3 SiO$_2$ Nanoparticles

Silica is one of the most widely used materials because it has biocompatibility without toxicity. Silica-based nanoparticles are nonluminescent in nature. However, their luminescent characteristics can be enhanced when these particles are doped with organic dyes or rare earth compounds. A wide array of colored and fluorescent molecules such as basic red 28, basic yellow 40, fluorescein, methylene blue, oxazine perchlorate, rhodamine 6G (R-6G), rhodamine B and thiazole orange can be used for developing and detecting fingerprints. SiO$_2$ nanoparticles doped with these compounds can be used, as aqueous suspension as well as dusting powder. Additionally, the surface of silica shells can be treated by various organosilicon compounds, which develops the latent fingerprints easily. The use of silica nanoparticles (SiO$_2$ nanoparticles) serves as a novel method for improving the detection of latent fingerprints because of their ease of synthesis and their ability to coat dyes which prevents photodecomposition.

Hydrolysis-condensation reaction of tetraethyl orthosilicate is used as a precursor with either acidic or basic catalysts for the synthesis of silica shells. Modified surface of QDs by ligand exchange methods and QDs coated with amphiphilic polymers have to adapt a variety of applications [51, 56]. Ryu et al. have studied the latent fingerprint detection by using SiO$_2$ QDs as fluorescent nanomaterials. SiO$_2$ nanoparticles were prepared by a reverse micelle method of water in oil, which easily controls the aggregation of QDs in the core. They observed strong fluorescence that could efficiently make visible latent fingerprint on different substrates such as aluminum foil, paper, plastic and glass [61]. SiO$_2$-doped nanoparticles can be used, as aqueous suspension as well as dusting powder, to detect fresh and aged latent fingermarks. Functionalized silica nanoparticles can be used to determine the molecular composition of fingermarks. Positively charged silica nanoparticles and hydrophobic ones are used to separate polar components present in the secretion residue left on a glass substrate.

SiO$_2$ nanoparticles have potential to recognize the fingerprint residue due to coating of dye particles, and the outer shell of SiO$_2$ was functionalized in such a manner that the organic constituents of sweat residue like amino acids and proteins interact with SiO_2 nanoparticles on varied surface types [50].

The SiO_2 nanoparticles with an average diameter of 70 nm have an excellent ability to coat and are therefore used as an advanced technology in developing latent prints on nonporous surfaces [50]. Another similar technique involves carbon dots coated with fluorescent starch powder. It is seen to have better visualization and clearer ridge details in comparison to cyanoacrylate fuming or commercial titanium dioxide powder conjugates on nonporous surfaces [50]. In addition to this, the electrodeposition of metal nanoparticles may give more distinct ridge details. The silica nanoparticles having both hydrophobic and hydrophilic parts when coated with 4(chloromethyl) phenyl trichlorosilane shows greater binding affinity toward amino acids present in sweat. These new methods of latent fingerprint development are more advanced and yield better results in contrast to previously used conventional powders for latent print development.

Recent study by Jenie et al. presented that the silica nanoparticles were synthesized from geothermal power plant waste and its subsequent conversion into silica-based fluorescent nanoparticles by using R-6G [27]. The maximum surface area of SiO_2 nanoparticles was observed to be 289.2 m^2/g before modification, and its morphology and physical properties showed particle size of 5–10 nm after modification. This interaction between silica nanoparticles and the fluorescent R-6G dye is due to hydrogen bonding (as seen through Fourier transform infrared spectroscopy) [27]. The silica-based nanoparticles synthesized from geothermal wastes were found to develop good-quality latent prints on nonporous dry smooth surfaces. Also due to its low toxicity and enhanced results, it may be utilized in place of traditionally used fingerprint development powders. The issue of sensitivity and selectivity in biochemical sensing using terahertz technology has been resolved by using new sensing scheme with two identical photonic crystal slabs (both consisting of silicon-based square lattice on silicon substrate). The guided resonance peak at 529.2 GHz with a high-quality factor of 529 has been obtained by optimization of geometric parameters of the cavity.

The effectiveness of the formulated cavity has been employed for the detection of lactose molecule few micron thicknesses, and it showed more than 30 times better in sensitivity with highly sensitive terahertz fingerprint sensing which were used to develop the latent fingerprints from scene of crime [8]. Rhodamine B-diatomaceous earth composites have been employed to develop latent prints due to physical adsorption between cationic fluorophores and porous silica. Researchers studied thoroughly on its sensitive development of latent fingerprints using rhodamine B-diatomaceous earth composites and principle of efficient image enhancement [79]. Using this technology, second- and third-level information of fingerprints were obtained. This explains the sensitivity and preciseness of the dye used in developing latent prints. On visualizing these prints in luminescence mode enhances the ridge details while diminishing the background noise. This technique may ideally be used in forensic applications of silica-based rhodamine B-diatomaceous earth composites to develop latent prints in real crime scenarios [79]. Recently, Tang et al. developed silica nanoparticles dually emissive in aqueous dispersion with two emissions [64]. The fluorescence emissions with various wavelengths from green to red region Si nanoparticles yield independently to various solvents with wavelength range from green to red of fluorescence emissions. However, the emission color and the fluorescence wavelength can both be predetermined or may be regulated as per the solvent system.

A sequence of various fluorescent (emission) spectra can be obtained using different solvents, for example, with copper ions it gives spectra of wide range of selection from green to blue (254–436 nm wavelengths). The prominent sensitivity of Si NP-A toward Cu^{2+} has been found with low detection limit and showed best fit linear relation between the ratio of fluorescence intensity (I_{436}/I_{500}) and concentration of Cu^{2+} [64]. The Si NP-A was employed to visualize the latent fingerprints due to dual-fluorescence visualization agent and feature of dual emission [64]. The latent prints obtained using this technique may reveal second-level ridge details which are useful in comparison and matching of crime scene prints with those of suspect prints.

4.3.2.4 Titanium dioxide (TiO$_2$) Nanoparticle

TiO_2-based nanoparticles have been used to enhance patent ridge detail found on relatively smooth dark surfaces. The reagent is a suspension of TiO_2 in a carrier liquid. The most common use for this reagent at scenes is for the enhancement of patterns of dried blood. Enhancement occurs through the adherence of TiO_2 to blood stuck to the smooth surface. The white color of TiO_2 makes it appropriate for enhancement of patterns observed on relatively smooth dark surfaces, where it provides additional contrast. The methanol-based solution is preferred when the surface in question is suitable. TiO_2 powder is used in the development of latent fingerprint on dark surfaces and night scene. If we are using a substitute for sticky-side powder, they produce excellent finding on black electrical tape and develop prints on both sides of the tape as well as on plastic bags and cellophane. TiO_2 powder may be used as either white fingerprint powder or white SPR. The TiO_2 is suspended in a surfactant or mixed in water is called as SPR which employed to analyze the prints on adhesive tapes. Scanning electron microscopy (SEM) is a reliable technique to show a range of levels of aggregation of particles on images of prints developed with different powders [10, 54]. Another study that deals with aging of fingerprints over a period and its relevance is in forensic case works.

The comparison of flat and rolled inked impressions was sought to ascertain whether the prints were given in recent days (fresh prints) or have been collected after some time elapsed since the commissioning of crime (aged/old prints). The study revealed correlation between fingerprints developed by TiO_2-based powders and traditionally used carbon black powder on a small study group of about 10 subjects. The results of this study showed a significant difference between flat and rolled prints as well as between ridge widths of latent and inked prints [55]. Comparable width ridges were observed in latent print with CB and TiO_2 powders [14]. The character of the charge carriers can be explored by using transient absorption spectroscopy but carrying out experiments on well-defined nanoparticles with a given morphology and selected size is extremely difficult [48].

In a study by Morales-García et al. [48], a hybrid time-dependent density functional theory–based calculation was carried out for realistic TiO_2 nanoparticles with bipyramidal, truncated and spherical morphologies. This revealed that the electron-trapped carriers are sensitive to nanoparticle morphology [2, 48]. The results showed that TiO_2 nanoparticles can also be combined with ZnO or WO_3 nanoparticles (or any other photoactive metal oxide) to extract information about molecular structure and bonding between nanoparticles morphology and the nature of the low-lying excited states [48]. It was suggested that the information is helpful for understanding the relation between structural and morphological properties and to improve the performance as a photocatalysis. The use of DFO (1,8-diazafluoren-9-one) is widely known in forensic applications, especially in developing latent fingerprints. DFO reacts with amino acids present in sweat to give fluorescent visible prints and can be used in combination with other methods such as ninhydrin method of developing latent prints from crime scene porous surfaces. The fluorescence detection of glycine (α-amino acid) was possible to immobilize DFO into a TiO_2 matrix [33]. The stabilization of DFO on a transparent substrate allowed its complex formation with biomolecules for the detection of α-amino acids. The same can be utilized to visualize latent prints as it adheres to the amino acids of sweat by forming DFO–TiO_2 optical thin film as shown in Figure 4.8.

4.3.2.5 Zinc Oxide (ZnO) Nanoparticles

ZnO is a popular semiconductor having a band gap of 3.37 eV. It has been found in three crystal structures: hexagonal wurtzite, cubic zinc blende and cubic rock salt. In general, hexagonal phase is relatively most stable at room temperature. High chemical stability, electrochemical coupling coefficient, photostability and broad range of radiation absorption make it useful for different operations [43].

It has good transition even at room temperature due to high excitation binding energy (60 MeV) as well as an extra binding property that permits contact of fingerprint residues such as lipids and proteins with powder for developing the latent fingerprints on nonporous surfaces. Consequently, clear ridge characteristics have been observed under UV radiation. It has been used to develop clear prints since it is naturally fluoresced in UV radiation under wet conditions. Also, it produces good fluorescent image of the latent fingerprints when illuminated by near-UV visible light. A new combination of ZnO–SiO_2 nanoparticles in powder form has been observed to be effective for the development of the latent prints on various nonporous surfaces. ZnO nanoparticles are less luminescent than other conventional commercial

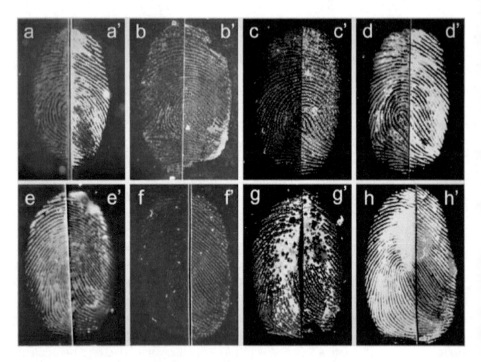

Figure 4.8: The luminescence of 1,8-DFO-9-one/TiO$_2$ composite thin films [33].

powders. However, these particles showed excellent ridge detail with minimum background staining. Due to its white color, it is frequently used in small particle reagent formulations to detect latent fingermarks on black and dark-colored nonporous surfaces. ZnO, as a fluorescent pigment, can be applied as a dry powder dusting or as small particle reagent formulation to detect latent fingermarks on a wide range of nonporous surfaces of forensic importance (Wang et al., 2015c). However, the use of ZnO nanoparticles as small particle reagent formulation is more useful and effective than the dry powder dusting because background staining on some surfaces, especially polyethylene, is observed with the latter method. Sometimes, before its application, particles of ZnO are mechanically grounded to ensure its size in the range of nanoparticles.

ZnO nanoparticles were utilized as a nanopowder for building up the matured inactive fingerprints on nonporous surfaces. Fluorescent N-carbon dots/ZnO nanopowder as powder dusting method has been used to be effective for development of latent print [53]. A new approach of using nitrogen-functionalized carbon dots coated on ZnO nanoparticles is a novel nanocomposite for latent print development. Melamine, potato peel waste and zinc acetate dehydrate as precursors have been used to synthesize and were characterized by many spectroscopies and microscopic techniques. The size of N-carbon dots and ZnO nanoparticles was around 50–20 nm and 40–50 nm, respectively [53]. To increase the quantum yield and blue emission,

N-carbon dots were coated on surface of ZnO nanoparticles. The novel nitrogen-functionalized carbon dots coated on ZnO nanoparticles (N-carbon dots/ZnO nanoparticles) were able to provide the second- and third-level ridge details on developed latent prints [53].

The formulation was seen to be highly sensitive and worked best on an array of nonporous substrates such as Al-based objects (Al foil, Al sheet, Al rod), Fe disk, CD-Rom (read only), marble/pebbles (white and black colored) and paper (glossy magazine sheet) [53]. The prints developed were clear and sharp and with distinct ridges when observed under UV lamp. The second phases of unique mark edge designs were created on the aluminum foil surface utilizing carbon dots/ZnO nanocomposite under the UV light illumination at 365 nm, as shown in Figure 4.9. The subsequent-level edges are shown around the latent print picture and indicated the subsequent stage edge designs including whorl, bifurcation, delta, endpoint, fork, island, hybrid and extension, as shown in Figure 4.9. Moreover, it showed better capability to divulge ridge details and high contrast latent images at 415 and 450 nm wavelengths [53].

Carbon dots coated on ZnO nanoparticles represented enhanced efficacy, nonbiohazardous, better results in the latent print detection with optimum optical characteristics. Hence, it may be used as an alternative for detection of latent fingerprints in crime investigations [53]. Vacuum metal deposition (VMD) using consecutive Au and Zn affidavits has been a successful procedure to create latent fingerprint on plastic surfaces.

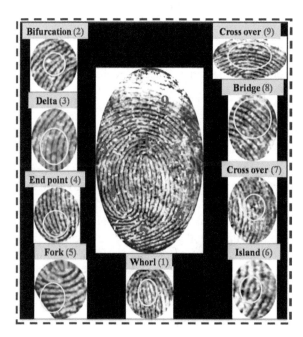

Figure 4.9: Dissimilar ridge patterns of latent print developed by applying N-cadmium sulfide/zinc oxide nanoparticles on Al foil under 365 nm ultraviolet light. Reproduced with reprint permission from [53].

In this technique, unadulterated ZnO is thermally vanished in a vacuum framework; thus, it can consolidate on polyethylene terephthalate surface. The improvement of latent finger impression by ZnO application is more efficient than that by Au/Zn VMD [25]. It can be utilized for the representation of latent fingerprints under ordinary light on different permeable and nonpermeable surfaces.

4.3.2.6 Al_2O_3 Nanoparticles

A nontoxic, ecofriendly nanoparticles were synthesized by using ecofriendly binders such as aluminum oxide (Al_2O_3) nanoparticles coated with a naturally colored dye (eosin yellow) and hydrophobic seed extract for various purposes. Nanopowders were not only used as a developing powder of latent print and providing sharp visibility on various substrates but also identifies the faint latent prints. The rare earth-doped composites such as luminescent materials have received remarkable identification worldwide. Al_2O_3 is a well-known host matrix for rare earth ions due to its high melting point, chemical inertness and photochemical stability [76]. Formulated rare earth ion Eu^{3+}-doped α-Al_2O_3 nanocrystalline powder is used for developing latent fingerprints. The structural and morphological characterization and phase identification can be done by X-ray diffractometry and field-emission scanning electron microscope. In order to visualize latent print under UV radiation, Al_2O_3:Eu^{3+} nanopowder can be used on various surfaces. The prints display good contrast on different substrates such as stainless steel plate, aluminum foil, glass slide, stainless steel bowl and highlighter pen [76]. Therefore, Al_2O_3:Eu^{3+} nanocomposite has the ability to develop latent prints, since it gives bright luminescence under UV radiation [76].

4.4 Rare Earth Materials

Rare earth element are a set of 17 elements in the periodic table, explicitly the 15 lanthanides, just as scandium and yttrium. Rare earth elements belong to sixth and seventh group elements of periodic table along with lanthanum and actinium. It is divided into two series, namely lanthanide and actinide series. The lanthanide series elements have been gained much attention over the past decades due to their unique optical properties. The lanthanide series elements have been devoted much consideration over the last decades due to their spectral and optical properties. Rare earth elements are filled 4f shell, shielded by $5S^2$ and $5P^6$ electrons, which makes them sensitive and more reactive. When they are added with host materials, they exist as 3+ or 2+ ionic states. Rare earth–doped nanomaterials significantly change the optoelectronic and spectroscopic properties of materials. Thus, these materials have gained much attention as a phosphor material since the last two decades [45, 68–72].

Rare earth elements are found as oxides, metals and fluorides. Due to their shielded 4f shell and f–f intra, parity forbidden transitions, these materials exhibit stable and sharp emission and are widely used as phosphor materials [45, 72].

4.4.1 Rare Earth Luminescent Materials for Development of Latent Prints

Rare earth luminescent materials emit the shorter wavelength light with higher wavelength excitation. For instance, when excited by near-infrared (NIR) radiation, these rare earth materials emit visible light. They show low toxicity, narrow emission bands and strong intensity of light as compared to transition metals. Moreover, rare earth luminescent materials can be used chemically and lead to a well-developed contrast. Therefore, they have been incorporated for developing latent print exhibiting high sensitivity, sharp contrast and enhanced selectivity. Presently, $NaYF_4$ co-doped with Yb^{3+} and Er^{3+} ions ($NaYF_4$:Yb,Er) have been used as upconversion fluorescent materials [70]. In this section, we discuss the application of various rare earth fluorescent materials for latent fingerprint development and detection through powder dusting method and liquid media method.

4.4.2 Dusting Technique

Doped rare earth materials and nanomaterials are systematically used to develop print residues using powder ducting methods as they stick to the greasy substance to facilitate the development on different nonporous and semiporous substrates [66]. Another research reported the use of $NaYF_4$:Yb,Er to develop fingerprints. This process is quite alike the conventionally used powder dusting method using squirrel's brush. Some impressive results on polymer banknotes were obtained, which could show intense fluorescence under UV radiation to lead a contrast background [70]. Similar results were reported with the use of YVO_4:Yb,Er bulk size for developing latent fingerprints on various nonporous and semiporous substrates [43, 44], although $NaYF_4$:Yb,Er and YVO_4:Yb,Er powders have been applied with micrometer size regime. It opened up a new possibility on the use of nanosized upconversion fluorescent particles for latent fingerprint development. $NaYF_4$:Yb,Er has been extensively used in labeling novel fluorescent print development substrates. These have been found to be actively working in any nonporous surface type with strong fluorescence, contrast and selectivity [70, 72].

Wang et al. used NaYF$_4$:Yb,Er nanoparticles to develop latent prints on an array of surfaces (Wang et al., 2015b). Solvothermal approach was used using oleic acid surface modifier for Yb,Er rare earth–doped NaYF$_4$ samples. Then, NaYF$_4$:Yb,Er powder samples were precisely sprinkled over the latent prints with the help of squirrel brush. The prints were then viewed under 980 nm NIR radiation after successful development (Wang et al., 2015a). The upconversion rare earth–doped NaYF$_4$ nanomaterial sample has proved significant improvement in developing prints on glass substrates as well (Wang et al., 2015a, Wang et al., 2015b). The developed prints showed clear and distinct ridge characteristics as well as sweat pores. On viewing the developed prints under NIR, a distinct green fluorescence could be observed at the dermal ridges [45, 69, 70].

Keywords: rare earth, dusting, fluorescent, substrate, fingerprint

4.4.3 Liquid Process

Rare earth upconversion nanocrystals can also be useful in colloidal or suspension form for latent print development and detection. In this regard, Wang and coauthors successfully used lysozyme-binding aptamer (LBA)-modified NaYF$_4$:Yb,Er rare earth–doped nanomaterials (NaYF$_4$:Yb,Er/LBA) for latent print investigations. The use of LBA-modified NaYF$_4$:Yb,Er nanomaterials for developing latent fingerprints was based on the molecular recognition technique [70]. DNA aptamer LBA targets lysozyme and identifies lysozyme, one of the polypeptides found in human perspiration [76]. In the reported process, firstly a suspension of NaYF$_4$:Yb,Er/LBA rare earth nanomaterials was employed to the latent fingerprints, and then the samples were incubated for half an hour. During incubation, the lysozyme molecules were selectively conjugated with LBA molecules on NaYF$_4$:Yb,Er surface [70]. The fingerprint residue consists of genetic markers specific to personal identification of individuals. This triggers the binding between rare earth nanomaterials NaYF$_4$:Yb,Er and latent fingerprints as a result of selective recognition and adsorption [70]. The surfaces that were treated with CdTe QDs exhibited no visible prints except only high violet fluorescence at the background was observed. However, under 365 nm UV excitation, the suspension of NaYF$_4$:Yb,Er showed prominent fluorescence.

Fluorescent NaYF4:Yb,Er are rare earth nanomaterials that were used for incubation, a clear fluorescent image without any background fluorescent interference was obtained. Also, the identification of arches and termination points of fingerprints had easily detectable ridge details when viewing under microscope. Although a relatively longer time of incubation (30 min) is required for this strategy, it works as a robust technique to develop latent fingerprints [70]. Rare earth fluorescent material–based suspensions for developing latent fingerprints were employed along with a solution of sodium dodecyl sulfonate (SDS) acting as a surfactant in 100 mL of distilled water.

The procedure utilized washing of latent prints with suspension mixture of SDS and Rare earth florescence materials for several minutes in a gentle stream of water. The hydrophobic end of the SPR attaches to greasy residue present in sweat [70]. An excitation with 980 nm NIR radiations enabled the detection of fingerprints. This method is also applicable to develop old, wet, autofluorescent and colorful background [45, 72]. Two types of rare earth fluorescent nanomaterials, including red emitting YVO_4:Eu nanocrystals and green-radiating $LaPO_4$:Ce,Tb nanowires, were combined by means of an aqueous media, as powerful fluorescent characteristics in developing the latent fingerprints. The fluorescent nanomaterials were then effectively utilized as feasible fluorescent dyes [70]. Latent fingerprints have been developed successfully on different substrates such as glass, artistic tiles, marbles, aluminum combination sheets, polymer materials, wood materials and papers (shown in Figures 4.10 and 4.11) [70, 72].

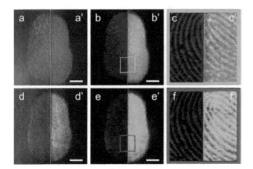

Figure 4.10: Latent fingermark on glass marked by YVO_4:Eu (a–c) and Ce- and Tb-doped $LaPO_4$; (d–f) fluorescent powders, under 254 nm irradiation, and without 254 nm UV light. The scale bar relates to 5.0 mm. Reprint permission from Wang et al. (2015).

The advanced nanomaterials represented wonderful results with high affectability, peculiarity and selectivity. Hence, the rare earth nanomaterials show promising opportunity for feasible applications in legal sciences (Wang et al., 2015b, Liu et al., 2015a and Liu et al., 2015b, [45]).

Another recent study reported the green and red color emission from europium- and terbium-doped $CaMoO_4$ nanoparticles in aqueous medium and in using Isopropyl alcohol IPA solvent [5]. They used strong red and green emissions of rare earth–doped $CaMoO_4$ for the advancement of latent fingerprint detection [5]. The observations on different surface types demonstrated sharp contrast with minimum background intrusion. Figure 4.12 depicts the visualized fingerprint on various surfaces by europium-

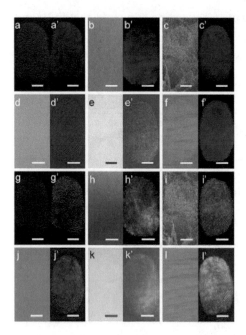

Figure 4.11: Latent fingermark on different surfaces marked by YVO$_4$:Eu (a–f) and LaPO$_4$:Ce,Tb (g–l) fluorescent nanomaterials in splendid field, and afterward distinguished by 254 nm UV light in dim field (a′– l′): (an, a′, g, g′) aluminum combination sheets, (b, b′, h, h′) artistic tiles, (c, c′, l, l′) marbles, (d, d′, j, j′) painted wood, (e, e′, k, k′) printing papers and (f, f′, l, l′) wood floor. The scale bar relates to 5.0 mm. Reproduced with reprint permission from Wang et al., 2015a [45], [68] and [69].

and terbium-doped CaMoO$_4$ phosphors. The significant improvement in the visibility of images under the UV light indicates the usefulness of CaMoO$_4$ phosphor materials. Further study also confirms that rare earth–doped CaMoO$_4$ can offer a better choice for fingerprint detection and development [13, 66]. Moreover, the fresh fingerprint image and powder-smeared images were visibly evident because of binding of nanoparticles to the moisture and oily constituents present in sweat residue present in latent. All the minute details like endings of ridges (Figure 4.13) evidently permit these nanomaterials as a potential candidate for latent fingerprint detection [5]. The pictures of the latent fingerprint (Figure 4.14) clearly indicate the difference between ridges and furrows and corresponding fluctuations. The fingerprints found on stainless steel and CD were preserved in ambient conditions for 9 days to test the quality of the powder and its affinity characteristics [5]. Hence, the observation using CaMoO$_4$ nanoparticles to develop latent fingerprints exposed successful ridge details up to third level of personal identification. Moreover, reported results suggest that these materials can also be used as an efficient alternative method for development and enhancement of latent fingerprints on various nonporous surfaces such as glass

Chapter 4 Nanophosphors: Emerging Materials for Forensic Applications — 155

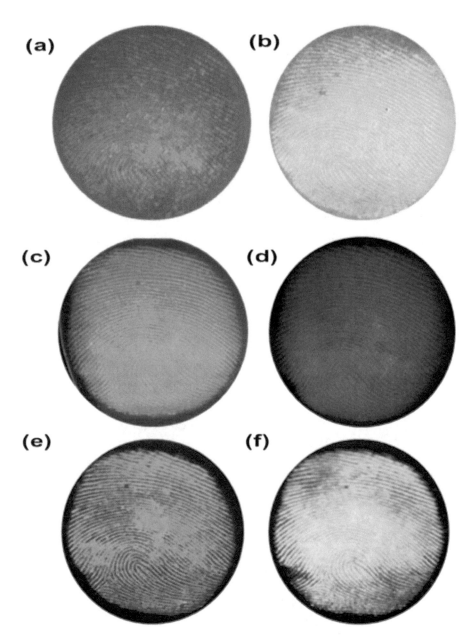

Figure 4.12: Digital pictures of the envisaged Eu-doped CaMoO$_4$ and Tb-doped CaMoO$_4$: on glass, CD and stainless steel (a, b; c, d; e, f), respectively. Reproduced with permission from [5].

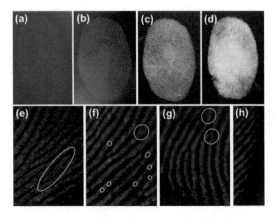

Figure 4.13: (a–f) Bare, powdered, fluorescent and high-magnification pictures of the latent mark under ultraviolet irradiation on stainless steel substrate. Reproduced with reprint permission from [5].

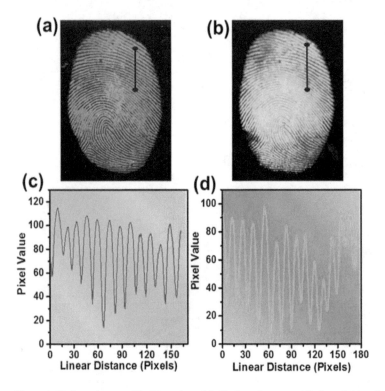

Figure 4.14: Europium- and terbium-doped $CaMoO_4$ spheres and their matching fluctuations. Reproduced with permission from [5].

slides, compact disks and stainless steel cups [5]. High contrast with poor background interference pictures were able to find minute details that aided in identification. Moreover, the good affinity of the powder presented an obvious fingerprint on aging (9 days) samples in ambient atmosphere. Therefore, the results are evident to show the characteristics of $CaMoO_4$ phosphor and the multipurpose nature of powders for latent print development [5].

4.5 Other Nanophosphors

Inorganic Eu^{3+}-doped α-Al_2O_3 nanocrystalline powder has good luminescent properties as well as thermal stability. Aluminum-based rare earth–doped nanophosphor has been used to develop latent fingerprints on various surfaces under UV excitation. It can have the ability to replace generally used luminescent materials, since it exhibits intense luminescence, which is certain requirement for quality fingerprint development as well as essential for minutiae ridge detail for fingerprint matching in forensic applications. It represented low interference background for studying the developed fingerprints [13].The better-quality performance of Eu^{3+}-doped $Y_4Al_2O_9$ phosphors over other bulk corresponding nanophosphor has shown to be industrial phosphor-converted light-emitting diodes (LEDs) and visualization of latent fingerprint [12]. In recent times, rare earth aluminate ($RAlO_3$) nanostructures are very important materials due to their excellent chemical, physical and optical properties and are useful for labeling agents for the visualization of latent prints which become important techniques for physical identification in criminology. Furthermore, the excellent green emission properties and the reliable result in latent print detection in green light visible region help for the investigation in forensic applications [28]. The exposure under UV, laser and ALS can be applied for developing fingerprint. Such methods explore the possibilities, since it fluoresce various surfaces.

Afterglow europium-doped strontium aluminate ($SrAl_2O_4$:Eu^{2+}) phosphor has demonstrated improved results for fingerprint development. It opened new possibilities. Strong afterglow effect of phosphor powder has become good candidates for better fingerprint detection. The standalone use of phosphor powder can also be used for developing latent fingerprints without any special device. The results yielded on various surfaces are reliable and prominent when developed fresh or even after few days. The bright afterglow is also applicable for cyanoacrylate fumed fingermarks. However, previous study represented that europium-doped strontium aluminate phosphor can also be used for the purposes of latent fingermark detection in powder form since it eliminates background completely [36]. Thus, rare earth–doped phosphor as the dusting agent has been used for latent fingerprint detection. It was able to detect unfumed as well as cyanoacrylate fumed fingermarks on various objects including metal foil, glass, porcelain, a plastic bag, paper, raw wood and fabric

[36]. Such phosphor allows strong and effective phosphorescence. Thus, $SrAl_2O_4:Eu^{2+}$ phosphors represent a new and useful class of phosphor powder for latent fingerprint detection [36]. Another Y_2O_3 phosphor has also been used for fingerprinting examinations due to the fast synthesis, small size and intense luminescence [37]. In another study, aluminum oxide phosphor and magnetic dichromate were utilized for development of latent prints. The Al nanophosphor was sprinkled over the greasy print on nonporous substrate while magnetic dichromate was spread over the print using magnetic wand. The freshly made prints can be developed using almost any method while those that older prints or aged can be developed using nanophosphor powders. These powders help in developing better quality prints, even with time elapse as nanophosphors are particularly fine (nano) sized to get better adhered to the grease in sweat residue and allow for good coverage to visualize older, more decomposed prints [37]. The size of the phosphor powder is comparable to commercially available bichromate powders; however, it shows better adherence on the latent prints as compared to other conventional powders. The improved usability of phosphor powder is also accounted due to its workable lesser amount in comparison to the other powders. Coated phosphors can be used for multipurpose application because it is more easily dispersed in different liquid media. It can also be used as a dip coating technique rather than the dusting technique. In addition, coated particles can be altered to attach various compounds to the surface which can add additional benefits to the powders [11, 37, 57].

Researchers associated SPR based on fluorescent dyes to develop latent prints from dirt-stained wet surfaces and surfaces exposed to high temperatures using powder suspensions. The added advantages of these powder suspensions over conventional methods are their ability to detect latent prints with low sensitivity and high selectivity, excessive background contrast and overcoming difficult surfaces such as those accidentally wetted or burnt [16, 36, 60].

The narrow emission band optical properties of fluorescent rare earth nanomaterials have found its applicability in various fields of forensic science such as urban farming. The fluorescent Mn^{4+} ions doped Ba_2LaNbO_6 ($BLN:Mn^{4+}$) have been used to develop latent fingerprints. These were also found to have shown applicability in plant growth LEDs using a facile citric acid–assisted gel solvent route [52]. The red photoluminescence spectra at 685 nm wavelength under 351 and 510 nm excitations optimized $BLN:0.25Mn^{4+}$ nanophosphors demonstrated detailed third-level unique identification features of latent fingerprints. The further fluctuation of red value spectra on pixel profile corroborates a clear affinity of nanophosphors $BLN:0.25Mn^{4+}$ due to its brilliant photoluminescence and wide detailing of the latent print identification and investigation [52].

4.6 Limitations and Challenges in Nanophosphor Powders

Nanotechnology and nanomaterial-based research and development is picking up fast in today's times. So is the case in forensic sciences where nanophosphors have been employed in detecting latent fingerprints and tracing evidence [70]. However, some major issues such as high cost and restricted synthesis conditions make limitation and restriction on the application of nanophosphors for dusting purposes. In case of rare earth–doped phosphor, the host lattice components are relatively cheap, the rare earth metals are relatively expensive and abundant. On the contrary, one of the advantages of producing nanophosphors is that it has lower health implications as in wet chemistry procedures where nanomaterials are formed. More often than not, latent prints are carried out by the dusting process. The dusting process is ideally carried out in open area or near proper ventilation in the laboratory so as to minimize the risk of inhalation of reduced size hazardous nanoparticles to examiners carrying out the experiments. The dusting powder-coated phosphors storage is another problem.

The silica coating adhered to the latent prints in aqueous medium penetrates through the storage medium such as sealed polybags and glass jars amplifies the particle agglomeration, thereby making the powder less efficient in adhering to the prints that have no better details could be identified in the dusting process [57]. The selection of powders is essentially based on size and shape-tunable luminescent properties and further development of latent prints in lesser time period and with much less requisites [21].

4.7 Conclusion

This chapter has overviewed the applications of nanotechnology in forensic science. More specifically, we have shown that semiconductor nanomaterials offer suitable alternatives of the conventional methods of fingerprint detections. We comprised a concise outline of the most widely used semiconductor nanomaterials for fingerprint and latent fingerprint detection. This chapter briefly explained the recent achievement in latent fingerprint development of fluorescent dyes using semiconductor nanoparticles, including QDs and rare earth upconversion nanomaterials. This is followed by the widely used conventional methods of developing latent prints such as powder dusting and liquid media methods. Drawbacks of these conventionally used fingerprint developing methods are also presented. To avoid these problems, nanomaterials that are fluorescent in nature, especially QDs and upconversion rare earth nanomaterials, have come up as a recent trend in fluorescent probes for developing latent prints. Nowadays, more advanced and sophisticated techniques have emerged for developing latent prints. Hence, the selection of the optimum method maybe based upon factors such as likelihood of high developing contrast, sensitivity, selectivity and low toxicity. The method employed should be simple and effective in detailing even faint or smudged prints. And finally, the method deployed may be used on an array of substrates such as nonporous, smooth, nonfluorescent and multicolored surfaces.

References

[1] Algarra, M., Reis, A., Bobos, I., Campos, B. B., Jiménez-Jiménez, J., Moreno-Tost, R., Esteves Da Silva, J. C. (2012). Optical characterization of CdS quantum dots nanoparticles dispersed in clays. J Dispers Sci Technol, 33(8), 1139–1143.

[2] Alivisatos, A. P., Harris, A. L., Levinos, N. J., Steigerwald, M. L., Brus, L. E. (1988). Electronic states of semiconductor clusters: homogeneous and inhomogeneous broadening of the optical spectrum. J. Chem. Phys., 89, 4001–4011.

[3] Bawendi, M. G., Carroll, P. J., Wilson, W. L., Brus, L. E. (1992). Luminescence properties of CdSe quantum crystallites: resonance between interior and surface localized states. J. Chem. Phys., 96, 946–954.

[4] Becue, A., Moret, S., Champod, C., Margot, P. (2009). Use of quantum dots in aqueous solution to detect blood fingermarks on non-porous surfaces. Forensic Sci. Int., 191(1–3), 36–41.

[5] Bharat, L. K., Ramaraju, G. S., Yu, J. S. (2017). Red and green colors emitting spherical-shaped calcium molybdate nanophosphors for enhanced latent fingerprint detection. Sci. Rep., 7(1), 1–14.

[6] Chan, W. C., Nie, S. (1998). Quantum dot bioconjugates for ultrasensitive nonisotopic detection. Science, 281(5385), 2016–2018.

[7] Chandra, B. P., Chandra, V. K., Jha, P. (2014). Luminescence of II-VI Semiconductor Nanoparticles. Solid State Phenom., 222, 1–65.

[8] Cheng, W., Han, Z., Du, Y., Qin, J. (2019). Highly sensitive terahertz fingerprint sensing with high-Q guided resonance in photonic crystal cavity. Opt. Express, 27, 16071–16079.

[9] Choi, M. J., McDonagh, A. M., Maynard, P., Roux, C. (2008). Metal-containing nanoparticles and nano-structured particles in fingermark detection. Forensic Sci. Int., 179, 87–97.

[10] Choi, M. J., Smoother, T., Martin, A. A., McDonagh, A. M., Maynard, P. J., Lennard, C., Roux, C. (2007). Fluorescent TiO2 powders prepared using a new perylene diimide dye: applications in latent fingermark detection. Forensic Sci. Int., 173, 154–160.

[11] Darshan, G. P., Premkumar, H. B., Nagabhushana, H., Sharma, S. C., Prashanth, S. C., Prasad, B. D. (2016). Effective fingerprint recognition technique using doped yttrium aluminate nano phosphor material. J. Colloid Interface Sci., 464, 206–218.

[12] Das, A., Saha, S., Panigrahi, K., Ghorai, U. K., Chattopadhyay, K. K. (2019). Enhanced photoluminescence properties of low-dimensional Eu^{3+} activated Y$_4$Al$_2$O$_9$ phosphor compared to bulk for solid-state lighting applications and latent fingerprint detection-based forensic applications. Microsc. Microanal., 25(6), 1422–1430.

[13] Das, A., Shama, V. (2016). Synthesis and characterization of Eu^{3+} doped α-Al$_2$O$_3$ nanocrystalline powder for novel application in latent fingerprint development. Adv. Mater. Lett., 7, 302–306.

[14] De, A. F. J., Tully, -. D. R., Weber, A. R., Barrot, F. C., Zapico, S. C., Rivera, C. N., Sirard, M. J., Graber, R. P. (2020). A small population study on friction skin ridges: differences in ridge widths between latent and inked fingerprints. J. Forensic Sci., 65, 620–626.

[15] Deepthi, N. H., Basavaraj, R. B., Sharma, S. C., Revathi, J., Sreenivasa, R. S., Nagabhushana, H. (2018). Rapid visualization of fingerprints on various surfaces using ZnO superstructures prepared via simple combustion route. J. Sci. Adv. Mater. Dev., 3, 18–28.
[16] Dhall, K., Sodhi, G. S., Kapoor, A. (2013). KA novel method for the development of latent fingerprints recovered from arson simulation. Egyp. J. Forensic Sci., 3(4), 99–103.
[17] Dilag, J., Kobus, H., Ellis, A. V. (2009). Cadmium sulfide quantum dot/chitosan nanocomposites for latent fingermark detection. Forensic Sci. Int., 187(1–3), 97–102.
[18] Edgar, J. H. The science of fingerprints: classification and uses. United States Federal Bureau of Investigation, 2006.
[19] Fern E, M. J., Watt, A., Warner, J., Cooper, S., Heckenberg, N., Rubinsztein-Dunlop, H. (2003). "Inorganic surface passivation of PbS nanocrystals resulting in strong photoluminescent emission". Nanotechnology, 14, 991–997.
[20] Gao, F., Han, J., Lv, C., Wang, Q., Zhang, J., Li, Q., Bao, L., Li, X. (2012). Application of core-shell-structured CdTe/ SiO2 quantum dots synthesized via a facile solution method for improving latent fingerprint detection. J. Nanopart. Res., 14(10), 1191.
[21] Girish, K. M., Prashantha, S. C. (2017). Nagabhushana H Facile combustion based engineering of novel white light emitting Zn 2TiO 4: Dy3+ nanophosphors for display and forensic applications. J. Sci. Adv. Mater. Dev., 2(3), 360–370.
[22] Gupta, S. K., Sahu, M., Gosh, P. S., Tyagi, D., Saxena, M. K., Kadam, R. M. (2016). Energy transfer dynamics and luminescence properties of Eu3+ in CaMoO4 and SrMoO4. Dalton Trans., 44, 18957–18969.
[23] Gurusinghe, N. P. ZnS Investigation of optoelectronic properties of thin film n-type ZnS on p-type Si (thesis). 2008.
[24] Hanus, L. H., Sooklal, K., Murphy, C. J., Ploehn, H. J. (2000). Aggregation kinetics of dendrimer-stabilized CdS nanoclusters. Langmuir, 16, 2621–2626.
[25] Yu, I.-H., Jou, S., Chen, C.-M., Wang, K.-C., Pang, L.-J., Liao, J. S. (2011). Development of latent fingerprint by ZnO deposition. Forensic Sci. Int., 20, 14–18.
[26] Huang, R., Liu, R. (2017). Efficient in situ growth of platinum nanoclusters on the surface of Fe3O4 for the detection of latent fingermarks. J. Mater. Sci., 23.
[27] Jenie, A. S., Krismastuti, F. S., Ningrum, Y. P., Kristiani, A., Yuniati, M. D., Astuti, W., Petrus, H. T. (2020). Geothermal silica-based fluorescent nanoparticles for the visualization of latent fingerprints. Mater. Express, 10, 258–266.
[28] Jisha, P. K., Prashantha, S. C., Nagabhushana, H. (2017). Luminescent properties of Tb doped gadolinium aluminate nanophosphors for display and forensic applications. J. Sci. Adv. Mater. Dev., 2, 437–444.
[29] John, P. Black, CLPE, CFWE, CSCSA, Senior Consultant, Ron Smith and Associates. Fingerprint training manual, federal bureau of investigation criminal justice information services division, identification services section, December 1993.
[30] Jin, Y. J., Luo, Y. J., Li, G. P., Li, J., Wang, Y. F., Yang, R. Q., Lu, W. T. (2008). Application of photoluminescent CdS/PAMAM nanocomposites in fingerprint detection. Forensic Sci. Int., 179(1), 34–38.
[31] Kitai, A. H. Luminescent Materials and Applications. John Wiley & Sons, Ltd, 2008.
[32] Lemon, B. I., Crooks, R. M. (2000). Preparation and characterization of dendrimer encapsulated CdS semiconductor quantum dots. Am. Chem. Soc., 122, 12886–12887.
[33] Lewkowicz, A., Bogdanowicz, R., Bojarski, P., Pierpaoli, M., Gryczyński, I., Synak, A., Mońka, M., Karczewski, J., Struck-Lewicka, W., Wawrzyniak, R., Markuszewski, M. J. (2020). The luminescence of 1, 8-diazafluoren-9-one/titanium dioxide composite thin films for optical application. Materials, 13, 3014.

[34] Li, J. J., Wang, Y. A., Guo, W., Keay, J. C., Mishima, T. D., Johnson, M. B., Peng, X. (2003). Large-scale synthesis of nearly monodisperse CdSe/CdS core/shell nanocrystals using air-stable reagents via successive ion layer adsorption and reaction. J. Am. Chem. Soc., 125(41), 12567–12575.

[35] Li, X., Wu, Y., Steel, D., Gammon, D., Stievater, T. H., Katzer, D. S., . . . Sham, L. J. (2003). An all-optical quantum gate in a semiconductor quantum dot. Science, 301(5634), 809–811.

[36] Liu, L., Zhang, Z., Zhang, L., Zhai, Y. (2009). The effectiveness of strong afterglow phosphor powder in the detection of fingermarks. Forensic Sci. Int., 183, 45–49.

[37] Liu, W., Wang, Y., Zhang, M., Zheng, Y. (2013). Synthesis of Y_2O_3: Eu^{3+} coated Y_2O_3 phosphors by urea-assisted homogeneous precipitation and its Photoluminescence properties. Mater. Lett., 96, 42–44.

[38] Liu, X., Li, L., Noh, H. M., Park, S. H., Jeong, J. H., Yang, H. K., Jang, K., Shin, D. S. (2015). Synthesis and photoluminescence of novel 3D flower-like $CaMoO_4$ architectures hierarchically self-assembled with tetragonal bipyramid nanocrystals. Opt. Mater., 43, 10–17.

[39] Lodha, A. S., Pandya, A., Shukla, R. K. (2016). The Nanotechnology: an applied and robust approach for forensic investigation. Forensic. Res. Criminol. Int. J., 2(1), 2–4.

[40] Luo, Y. S., Dai, X. J., Zhang, W. D., Yang, Y., Sun, C. Q., Fu, S. Y. (2010). Controllable synthesis and luminescent properties of novel erythrocyte-like $CaMoO_4$ hierarchical nanostructures via a simple surfactant-free hydrothermal route. Dalton Trans., 39(9), 2226–2231.

[41] Lv, L., Liu, X., Xu, C., Wang, X. (2014). Synthesis and optimum luminescence of doughnut-shape undoped and doped $CaMoO_4$. J. Nanosci. Nanotechnol., 14, 3521–3526.

[42] Ma, R., Shimmon, R., McDonagh, A., Maynard, P., Lennard, C., Roux, C. (2012). Fingermark detection on non-porous and semi-porous surfaces using YVO_4: Er,Yb luminescent upconverting particles. Forensic Sci. Int., 217(1–3), 23–26.

[43] Ma, R. L., Bullock, E., Maynard, P., Reedy, B., Shimmon, R., Lennard, C., Roux, C., McDonagh, A. (2011). Forensic Sci. Int., 207, 145.

[44] Wang, M., Ming, L., Aoyang, Y., Jian, W., Mao, C. (2015). Rare earth fluorescent nanomaterials for enhanced development of latent fingerprints. ACS Appl. Mater. Interfaces, 7(51), 28110–28115.

[45] Menzel, E. R., Savoy, S. M., Ulvick, S. J., Cheng, K. H., Murdock, R. H., Sudduth, M. R. (2000). Photoluminescent semiconductor nanocrystals for fingerprint detection. J. Forensic Sci., 45 (3), 545–551.

[46] Menzel, E. R., Takatsu, M., Murdock, R. H., Bouldin, K., Cheng, K. H. (2000). Photoluminescent CdS/dendrimer nanocomposites for fingerprint detection. J. Forensic Sci., 45(4), 770–773.

[47] Morales-García, Á., Valero, R., Illas, F. (2020). Morphology of TiO_2 nanoparticles as fingerprint for the transient absorption spectra: implications for photocatalysis. J. Phys. Chem. C, 124(22), 11819–11824.

[48] Moret, S., Bécue, A., Champod, C. (2013). Cadmium-free quantum dots in aqueous solution: potential for fingermark detection, synthesis and an application to the detection of fingermarks in blood on non-porous surfaces. Forensic Sci. Int., 224(1–3), 101–110.

[49] Moret, S., Becue, A., Champod, C. (2014). Nanoparticles for fingermark detection: an insight into the reaction mechanism. Nanotechnology, 25, 425502.

[50] Ow, H., Larson, D. R., Srivastava, M., Baird, B. A., Webb, W. W., Wiesner, U. (2005). Bright and stable core– shell fluorescent silica nanoparticles. Nano Lett., 5, 113–117.

[51] Pavitra, E., Raju, G. S. R., Park, J. Y., Hussain, S. K., Chodankar, N. R., Rao, G. M., Huh, Y. S. (2019/2020). An efficient far-red emitting Ba_2LaNbO_6: Mn^{4+} nanophosphor for forensic latent fingerprint detection and horticulture lighting applications. Ceramics Int., 46, 9802-9809.

[52] Prabakaran, E., Pillay, K. (2020). Synthesis and characterization of fluorescent N-CDs/ZnO NPs nanocomposite for latent fingerprint detection by using powder brushing method. Arab. J. Chem., 13(2), 3817–3835.

[53] Prasad, V., Lukose, S., Agarwal, P., Prasad, L. (2020). Role of nanomaterials for forensic investigation and latent fingerprinting – a review, 65, 26–36.

[54] Prashant, V. K., Murakoshi, K., Yujiwada, Yanagida, S. (Editor) H. S. Nalwa. Nanostructured Materials & Nanotechnology. Academic Press, San Diego, 2002, 130.

[55] Qian, J., Zhang, C., Cao, X., Liu, S. (2010). Versatile immunosensor using a quantum dot coated silica nanosphere as a label for signal amplification. Anal. Chem., 82, 6422–6429.

[56] Reip, A. Studies on the Synthesis and Use of Rare Earth Doped Nanophosphors for Application on Latent Fingerprints. Wolfson Centre for Materials Processing Brunel University, 2015.

[57] Ramrakhiani, M., Gautam, N., Kushwaha, K., Sahare, S., Singh, P. (2014). Electroluminescence in chalcogenide nanocrystals and nanocomposites. Defect Diffus. Forum, 357, 127–169.

[58] Ramrakhiani, M., Nogriya, V. Synthesis and characterization of zinc sulfide nanocrystals and zinc sulfide/polyvinyl alcohol nanocomposites for luminescence applications. In: Polymer Processing and Characterization. Apple Academic Press, Toronto, NJ, 2012, 109–138.

[59] Rohatgi, R., Kapoor, A. K. (2016). Development of latent fingerprints on wet non-porous surfaces with SPR based on basic fuchsin dye. Egyp. J. Forensic Sci., 6(2), 179–184.

[60] Ryu, S. J., Jung, H. S., Lee, J. K. (2015). Latent fingerprint detection using semiconductor quantum dots as a fluorescent inorganic nanomaterial for forensic application. Bull. Korean Chem. Soc., 36, 2561–2564.

[61] Smith, A. M., Gao, X., Nie, S. (2004). Quantum dot nanocrystals for in vivo molecular and cellular imaging. Photochem. Photobiol., 80(3), 377–385.

[62] Sooklal, B. K., Hanus, L. H., Ploehn, H. J., Murphy, C. J. (1998). A blue-emitting CdS/dendrimer nanocomposite. Adv. Mater., 10, 1083–1087.

[63] Tang, M., Zhu, B., Qu, Y., Jin, Z., Bai, S., Chai, F., Chen, L., Wang, C., Qu, F. (2020). Fluorescent silicon nanoparticles as dually emissive probes for copper (II) and for visualization of latent fingerprints. Microchim. Acta, 187, 1–10.

[64] Taylor, M. K., Kaye, D. H., Busey, T., Gische, M., LaPorte, G., Aitken, C., Ballou, S. M., Butt, L., Champod, C., Charlton, D., Dror, I. E. (2012). Latent print examination and human factors: improving the practice through a systems approach, 7842.

[65] Tranquilin, R. L., Lovisa, L. X., Almeida, C. R. R., et al. (2019). Understanding the white-emitting CaMoO4 co-doped Eu3+,Tb3+, and Tm3+ phosphor through experiment and computation. J. Phys. Chem., 123, 18536–18550.

[66] Wang, J., Wei, T., Li, X., Zhang, B., Wang, J., Huang, C., Yuan, Q. (2014). Near-infrared-light-mediated imaging of latent fingerprints based on molecular recognition. Angew. Chem. Int. Ed., 53, 1616–1620.

[67] Wang, M., Zhu, Y., Mao, C. (2015). Synthesis of NIR-responsive NaYF:Yb,Er upconversion fluorescent nanoparticles using an optimized solvothermal method and their applications in enhanced development of latent fingerprints on various smooth substrates. Langmuir, 31(25), 7084–7090.

[68] Wang, M., Li, M., Yang, M., Zhang, X., Yu, A., Zhu, Y., Qiu, P., Mao, C. (2015). NIR-induced highly sensitive detection of latent finger-marks by NaYF4: Yb,Er upconversion nanoparticles in a dry powder state. Nano. Res., 8(6), 1800–1810.

[69] Wang, M., Li, M., Yu, A., Zhu, Y., Yang, M., Mao, C. (2017). Fluorescent nanomaterials for the development of latent fingerprints in forensic sciences. Adv. Funct. Mater., 27(14), 1606243.

[70] Wang, M. (2016). Latent fingermarks light up: facile development of latent fingermarks using NIR-responsive up-conversion fluorescent nanocrystals. RSC Adv., 6, 36264–36268.

[71] Wang, M., Li, M., Yu, A., Wu, J., Mao, C. (2015). Rare earth fluorescent nanomaterials for enhanced development of latent fingerprints. ACS Appl. Mater. Interfaces, 7(51), 28110–28115.

[72] Wang, Y., Wang, J., Ma, Q., et al.. (2018). Recent progress in background-free latent fingerprint imaging. Nano. Res., 11, 5499–5518.

[73] Wang, Y., Herron, N. (1991). Nanometer-sized semiconductor clusters: materials synthesis, quantum size effects, and photophysical properties. J. Phys. Chem., 95, 525–532.

[74] Wang, Z., Daemen, L. L., Zhao, Y., Zha, C. S., Downs, R. T., Wang, X., Wang, Z. L., Hemley, R. J. (2005). Morphology-tuned wurtzite-type ZnS nanobelts. Nat. Mater., 4(12), 922–927.

[75] Wang, X.-F., Peng, G. H., Li, N., Liang, Z. H., Wang, X. (2014). Hydrothermal synthesis and luminescence properties of 3D walnut-like CaMoO4: Eu3+red phosphors. J. Alloys Compd., 599, 102–107.

[76] Wood, M. A novel approach to latent fingermark detection using aptamer-based reagents (thesis). Sydney, Australia: University of Technology, Sydney, 2014.

[77] Xu, C., Zhou, R., He, W., Wu, L., Wu, P., Hou, X. (2014). Fast imaging of eccrine latent fingerprints with nontoxic Mn-Doped ZnS QDs. Anal. Chem., 86, 3279–3283.

[78] Yuan, C., Li, M., Wang, M., Zhang, X., Yin, Z., Song, K., Zhang, Z. (2020). Sensitive development of latent fingerprints using Rhodamine B-diatomaceous earth composites and principle of efficient image enhancement behind their fluorescence characteristics. Chem. Eng. J., 383, 123076.

[79] Zhang, L. Y., Fu, W. W., Zheng, G. H., Dai, Z. X., Zhu, Y. N., Mu, J. J. (2016). Morphology and luminescent properties of SrMoO4: Eu3+,Dy3+. J. Mater. Sci. Mater. Electron, 27, 5164–5174.

Kamal K. Kushwah, S. K. Mahobia, Vikas Mishra,
Sandeep Chhawra, Ratnesh Tiwari, Neha Dubey, Vikas Dubey

Chapter 5
Luminescence Studies of $Y_2Sr_3B_4O_{12}$ Phosphor Doped with Europium Ion

Abstract: Herein, synthesis and characterization of Eu^{3+}-activated $Y_2Sr_3B_4O_{12}$ (YSBO) phosphor was synthesized by the modified method. Effect of various doping ions inside the host YSBO is reported in photoluminescence studies. The prepared samples were characterized by X-ray diffraction analysis, and the crystallite size was calculated by the modified Scherrer formula. CIE (Commission internationale de l'éclairage) coordinate is used to determine the color purity of the prepared sample and it shows white light emission (touches white line boundary in CIE) for europium-doped phosphors. In this host matrix, europium ions give all possible transitions. So, the prepared phosphor may be useful for white light emission in several display devices.

5.1 Introduction

For white light emission, several phosphors are already reported by various authors A lot of research works have been carried out on Eu^{3+} [6, 7, 18] or Sm^{3+} [2, 4, 9, 13, 14, 16] ion–dopxed phosphors with varying host matrix for this purpose. But weak absorption near the blue region of these materials due to parity forbiddances and small optical oscillator strength makes those phosphors away from the industrial application. So, the suitable host matrix is in search for efficient and pure red emission. In this context, the $Y_2Sr_3B_4O_{12}$ (YSBO):Eu^{3+}-doped phosphor synthesized by the solid-state reaction method has been reported for the first time as per the available literature. The chapter describes the photoluminescence (PL) emission of YSBO:Eu^{3+}-doped phosphor for variable concentrations of doping ions. CIE (Commission internationale de l'éclairage) coordinates show the emission of light from the phosphor and are well depicted in the white light region.

Keywords: YSBO phosphor, europium doping ion, CIE coordinate, display application

5.2 Experiment

The starting materials used for sample preparation are Y_2O_3 (99.9%), Eu_2O_3 (99.99%), $SrCO_3$ (A.R.) and H_3BO_3 (A.R.). In the reaction process, some consumptions of B_2O_3 obtained by the decomposition of H_3BO_3 at 150–220 °C were observed over the melting temperature of 560 °C. Accordingly, in order to obtain YSBO:Eu^{3+}, 7% of excess H_3BO_3 was used. After intimately mixing the starting materials with ethanol, the mixture was heated at 750 °C, then calcined at 1,200 °C in air and cooled to room temperature. This is known as the modified solid-state method because in the previous method reported by Zhang et al. [16], was prepared the sample in the solid-state method [8, 12]. Stoichiometric calculations for sample preparation are:

$$Y_2O_3 + 3\,SrCO_3 + 4\,H_3BO_3 = Y_2Sr_3B_4O_{12} + 3\,CO_2 + 6\,H_2O$$

5.3 Results and Discussion

Figure 5.1 shows X-ray diffraction (XRD) pattern of YSBO phosphor doped with europium ions. Here, the XRD pattern is well matched with JCPDS card no. 48-0307 and shows the orthorhombic structure of strontium yttrium borate phosphor.

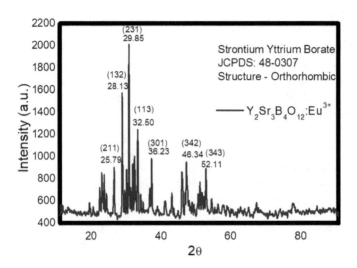

Figure 5.1: XRD pattern of $Y_2Sr_3B_4O_{12}$ (YSBO).

5.4 Photoluminescence Study

The PL emission spectra were recorded using 254 nm excitation and found various peaks of europium ions in the visible region such as the emission wavelength at 483 nm (blue region), 544 nm (green region), 594 nm (orange region) and 616 and 629 nm (red emission) due to $^5D_0 \rightarrow {}^7F_{0,1,2,3}$ transition (figure 5.2).

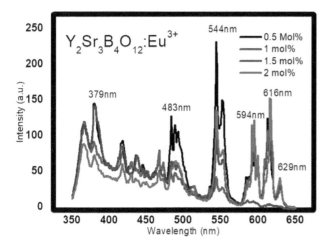

Figure 5.2: PL emission spectra of $Y_2Sr_3B_4O_{12}:Eu^{3+}$– doped phosphor.

It is visible from PL emission spectra that the intensity of red emission increases with increasing concentration of doping ion. Initially, the intensity of blue emission was high for lower doping concentration means that the effect of host material dominant the transition of dopant after increasing the europium concentration the dominant peaks are 616 and 629 nm (red emission) due to $^5D_0 \rightarrow {}^7F_{0,1,2,3}$ transition.

CIE coordinate shows the dominant peaks that are depicted inside PL spectra; here, the overall composition was found nearer to white light boundary in CIE coordinate for YSBO doped with Eu^{3+} phosphor ($X = 0.30$, $Y = 0.32$), which means that the prepared phosphor shows the white light emission inside the host YSBO matrix (figure 5.3).

Keywords: CIE coordinate, lower doping, emission spectra

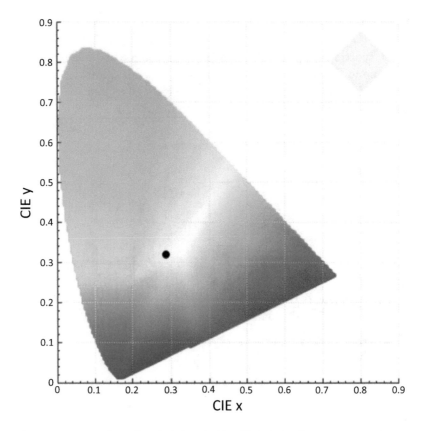

Figure 5.3: CIE coordinate for YSBO doped with Eu^{3+} phosphor (X = 0.30, Y = 0.32).

5.5 Conclusion

From this study, it is found that the prepared phosphor doped with europium ions shows well-resolved PL spectra in all visible regions and the dominant peak was blue for lower concentration; after increasing the concentration of dopant, the red emission peaks are dominant over the blue and other due to the transition of $^5D_0 \rightarrow {^7F_{0,1,2,3}}$. Hence, the prepared phosphor was useful for various display applications. The crystal structure was analyzed by XRD and it shows the orthorhombic structure of prepared phosphor.

References

[1] Li, X. X., Wang, Y. H., Chen, Z. (2008). Photoluminescence properties of GdBa3B9O18: Ln3+ (Ln= Eu, Tb) under UV and VUV excitation. J. Alloys Compd., 453(1–2), 392–394.

[2] Li, X. Z., Wang, C., Chen, X. L., Li, H., Jia, L. S., Wu, L., . . . Xu, Y. P. (2004). Syntheses, thermal stability, and structure determination of the novel isostructural RBa3B9O18 (R= Y, Pr, Nd, Sm, Eu, Gd, Tb, Dy, Ho, Er, Tm, Yb). Inorg. Chem., 43(26), 8555–8560.

[3] Liu, N., Zhao, D., Yu, L., Zheng, K., Qin, W. (2010). Controlled synthesis and photoluminescence of europium doped barium borate nanorods, nanowires, and flower-like assemblies. Colloids Surf. A Physicochem. Eng. Asp., 363(1–3), 124–129.

[4] Mo, F., Zhou, L., Pang, Q., Gong, F., Liang, Z. (2012). Potential red-emitting NaGd (MO4) 2: R (M= W, Mo, R= Eu^{3+}, Sm^{3+}, Bi^{3+}) phosphors for white light emitting diodes applications. Ceram. Int., 38(8), 6289–6294.

[5] Neeraj, S., Kijima, N., Cheetham, A. K. (2004). Novel red phosphors for solid-state lighting: the system NaM $(WO_4)_{2-x}$ $(MoO_4)_x$: Eu3+ (M= Gd, Y, Bi). Chem. Phys. Lett., 387(1–3), 2–6.

[6] Shinde, K. N., Dhoble, S. J. (2013). A novel reddish-orange phosphor, NaLi2PO4: Eu3+. Luminescence, 28(1), 93–96.

[7] Sohn, K. S., Park, D. H., Cho, S. H., Kwak, J. S., Kim, J. S. (2006). Computational evolutionary optimization of red phosphor for use in tricolor white LEDs. Chem. Mater., 18(7), 1768–1772.

[8] Sreebunpeng, K., Chewpraditkul, W., Nikl, M. (2014). Luminescence and scintillation properties of advanced Lu3Al5O12: Pr^{3+} single crystal scintillators. Radiat. Meas., 60, 42–45.

[9] Tian, L., Yu, B. Y., Pyun, C. H., Park, H. L., Mho, S. I. (2004). New red phosphors BaZr $(BO_3)_2$ and SrAl2B2O7 doped with Eu^{3+} for PDP applications. Solid State Commun., 129(1), 43–46.

[10] Wang, X. X., Wang, J., Shi, J. X., Su, Q., Gong, M. L. (2007). Intense red-emitting phosphors for LED solid-state lighting. Mater. Res. Bull., 42(9), 1669–1673.

[11] Wang, X. X., Xian, Y. L., Shi, J. X., Su, Q., Gong, M. L. (2007). The potential red emitting $Gd_{2-y}Eu_y(WO_4)_{3-x}$ $(MoO_4)_x$ phosphors for UV InGaN-based light-emitting diode. Mater. Sci. Eng. B, 140(1–2), 69–72.

[12] Wang, Y., Guo, X., Endo, T., Murakami, Y., Ushirozawa, M. (2004). Identification of charge transfer (CT) transition in (Gd, Y) BO3: Eu phosphor under 100–300 nm. J. Solid State Chem., 177(7), 2242–2248.

[13] Xiao, Q., Zhou, Q., Li, M. (2010). Synthesis and photoluminescence properties of Sm^{3+}-doped CaWO4 nanoparticles. J. Lumin., 130(6), 1092–1094.

[14] You, H., Wu, X., Zeng, X., Hong, G., Kim, C. H., Pyun, C. H., Park, C. H. (2001). Infrared spectra and VUV excitation properties of BaLnB9O16: Re (Ln= La, Gd; Re= Eu, Tb). Mater. Sci. Eng. B, 86(1), 11–14.

[15] Zaushitsyn, A. V., Mikhailin, V. V., Romanenko, A. Y., Khaikina, E. G., Basovich, O. M., Morozov, V. A., Lazoryak, B. I. (2005). Luminescence study of LiY_{1-x} Eu_x $(MoO_4)_2$. Inorg. Mat., 41(7), 766–770.

[16] Zhang, S., Zhu, B., Zhou, S., Qiu, J. (2009). Orange-red upconversion luminescence of Sm^{3+}-doped ZnO-B2O3-SiO2 glass by infrared femtosecond laser irradiation. J. Soc. Inf. Disp., 17(6), 507–510.

[17] Zhao, X., Wang, X., Chen, B., Meng, Q., Di, W., Ren, G., Yang, Y. (2007). Novel Eu^{3+}-doped red-emitting phosphor Gd2Mo3O9 for white-light-emitting-diodes (WLEDs) application. J. Alloys Compd., 433(1–2), 352–355.

[18] Zhou, Z., Wang, N., Zhou, N., He, Z., Liu, S., Liu, Y., . . . Hintzen, H. T. (2012). High colour purity single-phased full colour emitting white LED phosphor Sr2V2O7: Eu+. J. Phys. D: Appl. Phys., 46(3), 035104.

L. Castro-A., Ma. E. Zayas-S., Oscar R. Gomez-A., Julio C. Campos, C. Figueroa-N.

Chapter 6
Photoluminescence in Semiconductor Materials

Abstract: In this review work, the photoluminescence effect is presented by some semiconductor materials such as ZnS:Cu, CdTe and AuCl to mention a few, and their effect is generated by the recombination of the hollow electron pair that produces an exciton. The main motivation in research on the optical properties of materials is produced by changes in processes, excitation energy or variation in the concentrations of the reactants used. In the process of obtaining nanoparticles and doped thin films, techniques such as chemical vapor deposition, wet chemical at room temperature, the Czochralski method, sol–gel, chemical bath deposition were used. For the variation in the excitation energy, different intensities of laser light were used through monochromator spectrometers, and the variation in the percentage of doping of the materials was necessary. For the characterization of these, techniques such as X-ray diffraction were used in order to identify the formation of crystalline phases or the crystalline structure of the nanoparticles. Transmission electron microscopy allows us to know the size and distribution of the nanoparticles. Backscatter spectrophotometer helps us to determine the concentration of chemical compounds. atomic force microscopy is a surface technique, with which we know the roughness of thin films and spectrofluorometer with double excitation monochromators, emission and detectors of photomultiplier tubes, with which we obtain the emission and excitation spectra. The results obtained show that the variation in the doping concentration produces an increase in the size of the band gap in different doped materials, thus allowing an increase in the intensity of luminescence or increasing the amount of luminescent peaks at different wavelengths. The modification in morphology also produces changes in luminescence intensity when changing the scale from macromaterials to nanomaterials or the change in the roughness of thin films.

The study of these semiconductor materials and their research has allowed them to be incorporated into today's technology, increasing the efficiency of machines such as computers, telephones and in general the improvement of all electronic equipment that meets the needs of the population.

Keywords: photoluminescence in semiconductors, band gap, silicon, nanomaterials

6.1 Introduction

Luminescent phenomena have been studied for many decades [20, 8, 23, 24, 19]. The intensity of luminescence has an exponential dependent on the division of the quasi-Fermi level between electrons and holes and the open circuit, and the voltage can be tested, even before manufacturing an electrical device. Photoluminescence (PL) is a nondestructive technique [23, 7, 3] that has been widely used to study both radiative recombination at the edge of the bandgap and electronic levels within the bandgap band.

Figure (1) shows the shape of the temporal evolution in a study of PL in a material. II–VI semiconductors are materials [25] that have great importance in current technology due to their applications in optoelectronic devices. These materials [15, 5, 18, 23, 11, 1, 25] can be obtained in the form of thin films using techniques such as evaporation, sputtering, pyrolysis or chemical bath deposition. CdTe is a group II–VI semiconductor compound that is of special importance due to its technological applications, particularly in optoelectronic and photovoltaic devices and as a substrate for film deposition.

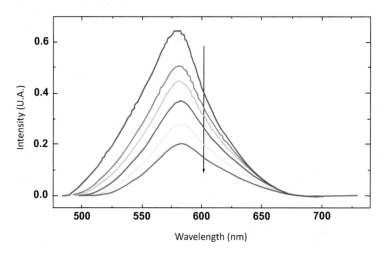

Figure 6.1: Photoluminescence spectra at room temperature of samples with semiconductors.

The remarkable property of porous silicon to present PL at room temperature, in the range of visible light when irradiated with ultraviolet light, has led to intense research activity. The high absorption coefficient and the PL and electroluminescence properties of porous silicon [2, 22, 10, 6, 13, 12] allow its application in the manufacture of optoelectronic devices.

The PL of graphene oxide is associated with the degree of oxidation of the graphene lattice, and the PL characteristics can be adjusted through alterations in the degree of oxidation. By oxidizing graphene [25] to form graphene oxide, the π electronic configuration of the carbon atoms in graphene is altered. This generates the development [11, 16] of a gap between the valence and conduction bands, which causes the material to become PL.

Nanocrystals are characterized by a dominant surface effect and have a considerable surface area/volume. In this sense, the optical properties of the semiconductor change considerably when their sizes are comparable to the Bohr radius of exciton. The characterization of AuCl nanocrystals in KCl has been studied through the use of X-ray diffraction (XRD) and PL [11].

Nanostructured oxides [2] have attracted great interest due to their unique properties and novel applications. Ion implantation is essentially a brute force (athermic) process, involving individual atoms, which is not restricted by law. The implanted ions [14, 21] can be placed at virtually any desirable depth and at any concentration in a solid (within the limitations of available ionic energy) without being restricted by diffusivity or solubility limits.

On the other hand, much effort has been devoted to the investigation of doped metallic chalcogenide nanostructures. These types of nanomaterials exhibit unusual physical and chemical properties compared to their bulk materials, such as the size-dependent variation of the energy band. These materials can produce unique magnetic and magneto-optical properties and provide unprecedented opportunities for the new field of spintronics (magnetoelectronics).

The objective of this work is the development of a bibliographic review on the subject of PL in semiconductor materials, knowing what has happened to this day and by the various research groups on this subject that belong to the different universities and research centers. It has been found that this topic is actually very broad and has applications in many types of materials, as well as various structures due to their dimensions.

Complementing this technique in carrying out studies of different phenomena with others has been a very interesting work for the group that we focus our attention on obtaining this work.

6.2 Experimental Procedure

PL is a well-established spectroscopic technique; its improvement and use have been observed in recent decades. Through the excitation with light to the materials, it has been possible to characterize them widely.

It is possible to mention case studies in the study of impurities, energy transfer between impurities, study of various defects and/or color centers and focusing our attention on semiconductors such as silicon, germanium and GaAs and various materials achieved from groups III to V of the periodic table.

In various studies established in this technique, it has been possible to obtain information on energy bands, forbidden bands or gaps, valence and conduction bands that in an interesting way explain the physics involved in the PL technique when a photon is absorbed by a material, and the formation of electron–hole pairs in the jump between valence and conduction bands. With the difference in energy between these bands, we can characterize the gap of these materials widely used today.

In order to obtain PL measurements in the laboratory in samples of semiconductor materials, the following points need to be considered:

(1) Having a lamp or source that emits electromagnetic radiation at a wavelength (λ) that does not present variants; at this point we must identify the power (watts) of the lamp.
(2) The appropriate optics in order to direct the beam of electromagnetic radiation toward the sample under study; in this case, the semiconductor can be specified in the appropriate lenses that meet this objective.
(3) A sample holder that supports it in order to place the sample as the experiment has designed and facilitates the study through the incidence of radiation to this special place of the study system.
(4) Appropriate lenses, prisms and optics to collect the response of the electromagnetic radiation (λ) arising from the sample.
(5) Within the individual instruments, a monochromator has the function of selecting the wavelength of radiation or light or obtaining the monochromatic light for a respective study.
(6) A device to detect the radiation signal emitted by the sample through the optics, and in turn, start the analysis process.

It is important to mention that the technology of incidence (radiation source and optics) or detection (optics, monochromator) of radiation emitted by the sample depends on the type of material under study. Similarly, the importance of the type of sample and the experimental conditions can be taken into account to select the incidence and data collection optics. The position and geometry of the sample are vital for the study; thus, the prior preparation of these is of great importance and a whole area of study.

From the information that we have about the energy intensity as a function of wavelength or frequency, it can be presented in 2D or x graphics, and that gives the luminescent information stimulated through light into the sample. The study can be carried out by scanning the x–y space of the sample or as an angular function in two-dimensional space.

With this technique, it is possible to scan the location of impure elements within the structure of the material under study, and each element shows a characteristic intensity versus wavelength graph structure, as well as to monitor the formation and growth of materials.

Being very observant and paying attention to the spectra obtained in the symmetry, central position of the graph, the FWHM of the peak, among others, can give important information for the study carried out in this optics technique.

Finally, PL was acquired in a spectrofluorometer using either a 450 W ozone-free Xe lamp, or a 325 nm He–Cd laser, coupled with a Horiba Triax 320 excitation and an emission iHR320 monochromator.

6.2.1 Czochralski Method

This method is widely used to obtain materials with monocrystalline arrangements, and from their melting they are grown in a rod with a rotating base in which a crystalline seed placed has a desired crystallographic direction, achieving very good quality crystals in a cylindrical and transparent way, followed by heat treatment and cooling process at a temperature and time that allows homogenization of the sample. For example, Kameyama et al. [22] used the benefits of this method and obtained samples of AuCl nanocrystals doped with KCl nanoparticles (NPs) with thicknesses of the order of 1 mm. This method has been widely used and was first proposed in 1916 by Jan Czochralski and is used at an industrial level to obtain semiconductor materials such as silicon.

6.2.2 X-Ray Analysis

X-ray analysis is a nondestructive analysis technique, which is useful for studying the crystalline structure through which the crystalline faces present in the material are identified and it reveals the information of the chemical composition. Identification of faces is performed by comparing it with a database already existing in the computer.

This technique is widely used to evaluate minerals, polymers, products with corrosion damage and unknown materials. The samples analyzed for their chemical elements are prepared using finely ground powders.

This test method is carried out by directing an X-ray beam at a sample and measuring the scattering intensity as a function of the exit direction. Through this technique, the identification of samples obtained by some method is facilitated, before proceeding to carry out our investigation. In a very recent study, Kameyama et al. [25] used this type of analysis in a study on Ag (In, Ga) Se$_2$.

6.3 Results and Discussion

Synthesis and PL of ZnS:Cu NPs are described [23]. In the XRD analysis of NPs of ZnS:Cu, the formation of the crystalline phase corresponding to the cubic Zn of ZnS is revealed. No copper peaks (such as CuS or CuO) were detected, at least within the resolution limit of the diffractometer. The magnification of the diffraction peaks of the NPs is obvious, which is characteristic of nanometric materials. The average size of these NPs from the high-resolution transmission electron microscopic observation is approximately 3.5 nm, where it is clearly observed that the particles have a crystalline structure.

On the other hand, the UV–visible absorption spectra of the ZnS:Cu NPs are exhibited, where an absorption shoulder was observed at approximately 316 nm (3.92 eV) and its position shows little change with the variation of the concentration of doping with Cu^{2+} (it must be mentioned in the synthesis). In the ZnS:Cu NPs, a green emission peak was observed at around 500 nm; the green emission arises from the recombination between the surface donor level (sulfur gap) and Cu^{2+} level 2.

It should be mentioned here that the prepared CuS NPs do not show any luminescence under UV radiation. Therefore, the observed green emission originates from Cu^{2+} ions that are embedded in the ZnS matrix and replace Zn^{2+} ions in ZnS NPs. With increasing Cu^{2+} concentration, the position of the green emission peak is systematically shifted to a longer wavelength (497–512 nm).

The results in [24] regarding the formation of SnO_2 NPs of the order of 30 nm embedded in $Al2O_3$, three SnO_2 emission bands are observed that reach a maximum of 410 nm, 520 nm and 700 nm and an absorption region in 280 nm. A PL band of the surface state of SnO_2 NPs that are embedded in Alumina has been observed.

The surface states located in the band gap trap electrons from the valence band to contribute to luminescence. However, the surface states [8] of SnO_2 NPs integrated in Al_2O_3 must be different from those of the NPs prepared by other synthetic methods.

Fig. 2 taken from the reference Rivera et al [19] a study of PL at room temperature can be observed, films of porous silicon (PS) of high PL [19] were formed by means of an anodic oxidation process. PL was carried out and the peaks were adjusted by Gaussian functions of their respective spectra, where they are presented at 1.8571, 1.8749, 1.8071 and 1.7331 eV for porous silicon prepared, respectively, at concentrations of 100%, 99%, 90% and 75% of HF in solution with ethanol (C_2H_5OH).

XRD of KCl-doped AuCl nanocrystals after annealing at 650 °C for 3 h. The spectrum shows the strong peaks of single KCl crystals [25] and two diffraction peaks with weak intensity, located at $2\theta = 31.74°$ and $2\theta = 65.83°$. These peaks are, respectively, attributed to planes [211] and [422] of the AuCl tetragonal structure, where AuCl has a preferential orientation along planes [211] and [422], due to the oriented growth of the crystals. Individual KCl is obtained by the Czochralski method. However, the spectrum of AuCl doped with KCl showed an emission band centered at 580 nm (2.14 eV). This band was assigned to the AuCl nanocrystals with a volume shift equal to 52 nm (0.18 eV). The widening of this band can be explained by the presence of AuCl nanocrystal size dispersion within KCl [24].

A calculation of the direct conversion of a measured luminescence signal [21] is developed from the Lasher–Stern–Wurfel (LSW) formula:

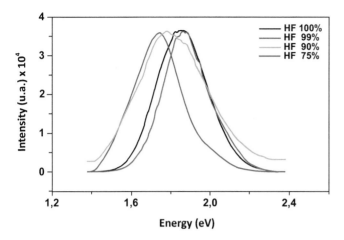

Figure 6.2: Photoluminescence spectra at room temperature of porous silicon samples prepared by anodic oxidation at concentrations of (a) 100%, (b) 99%, (c) 90% and (d) 75%, respectively, of HF in the mixture with C_2H_5OH (this was taken from [19]).

$$n(E)dE = \frac{a(E)\Omega}{4\pi^3 \left(\frac{h}{2\pi}\right)^3 c^2} \frac{E^2}{\exp\left(\frac{(E-\Delta E_f)}{k_b T}\right) - 1} dE$$

where

$$\alpha(E) = 1 - \exp[-a(E)d]$$

For an area A in the homogeneous case, the LSW formula already tells us the rate R of photons passing through an aperture with solid angle Ω per second, $R = A\, N(E)\, dE$. Multiplying this by P/P_{bare}, we have the total rate R reaching the detector, which has solid angle $\Omega = 2\pi(1 - \cos\theta_{obj})$,

$$R = \int \frac{P(E)}{P_{bare}(E)} \frac{A}{2\pi^2 \left(\frac{h}{2\pi}\right)^3 c^2} \frac{E^2}{\exp\left(\frac{(E-\Delta E_f)}{k_b T}\right) - 1} dE$$

R is numerically integrated over the energy range of expected luminescence, near the bandgap energy E_g.

In variable compositions, the complex index has been used in the range 1.1–1.3 eV of the PL energy spectrum of polycrystalline $CuIn_{1-x}Ga_xSe_2$ (CIGS) thin films [26] window, which was assumed to have a true refractive index of 2.3. The measured intensity of PL is equal to CR. Since C is established by calibration, the value of the quasi-Fermi level division ΔE_F that corresponds to the correct value for R is found in the equation.

The PL range of graphene oxide was clearly wide, spanning 350–1,250 nm. To calculate this broad spectrum of emission wavelengths, two photomultiplier tube detectors have been used: PMT-900 of the FLS1000 and PMT-1700 for the study intervals up to 1,700 nm. The PL was calculated using each detector, and the spectra were fused at 800 nm (the crossover point of the detector's quantum efficiencies), to provide the total spectrum. An FLS1000 can simultaneously house a maximum of five different detectors, and these can be easily changed using only Fluoracle software.

A study of a new type of PL properties [15] in multicolored organic materials of the crystals was observed by the combination of pure Terpy with impurities. Theoretical calculations [5] to obtain PL spectra using different methods are applied to nanostructured TiO_2 semiconducting materials in different quantum confinement regimes, showing optical gap narrowing due to confinement. The optical absorption and PL spectra of two synthetic diamonds in the temperature range of 80–300 K were measured. It is settled down that the free exciton state in a diamond is split by at least 12 sublevels. It has been observed in a study that I–III–VI2 semiconductor NPs are strong candidates for fluorescent materials composed of nontoxic elements substituting highly fluorescent CdSe NPs [1]. The study of NPs focuses on obtaining the PL spectrum of their defects [3] in materials with Cu [25]. The structural and optical properties of AuCl semiconductor nanocrystals embedded in KCl single crystals were investigated. The PL spectra presented a band emission situated at 580 nm, leading to the formation of AuCl nanocrystals within KCl. It has developed research [2] on the role of intrinsic defects in enhancing the performance of reduced TiO_2-based materials. Our in situ PL studies under controlled oxygen vacancy provides the PL information spectrum in the visible region, and other studies in Ag-(In-Ga)-Se(AlGaSe) quantum dots (QD) from visible to NIR (near infrared) [22]. In this study, it is important for designing and fabricating novel NIR-responsive devices with I–III–VI-based multinary semiconductor QDs. The PL performance and mechanism of nanosized semiconductor materials [10], such as TiO_2 and ZnO, are introduced, together with their attributes and affecting factors and can provide a firm foundation in theory for designing and synthesizing new semiconductor photocatalysts with high activity. The optical properties and the evolution of AgCl semiconductor aggregates dispersed in an ionic crystalline matrix, NaCl has also been investigated [14]. Several studies with PL in aggregated and doped materials have been carried out, as well as the dependence of PL with temperature and intensity of excitation [4] of the electrostatic potential in the epitaxial layers.

6.4 Conclusions

PL optical phenomena have been of great support in understanding the physical structure of various materials. In this work, it has been made possible to visualize a very broad panorama of works that focus our attention on a very rich area of research for future generations of students and researchers.

The result of this review allows us to interpret some characteristic phenomenologies of luminescence for various semiconductor materials obtained by the methods used such as chemical bath deposition, Czochralski method and method of vapor transport in closed space, being characterized by scanning electron microscopy, transmission electron microscopy, X-ray emission spectroscopy, atomic force microscopy, high-resolution electron microscopy and selected area electron diffraction, among other spectrometers such as double excitation and monochromators. In the studied samples, the authors presented the variations in concentration, and among other parameters that improve the properties of these materials, allow us to have a starting point to continue studying materials with these characteristics and thus get to know more about the best conditions, to work these and obtain better PL properties, as well as new applications for devices in industry and in health fields. In this work, we were able to discuss studies on various materials and theoretical calculations on the subject that we are convinced will be of great support.

References

[1] Hirase, A., Hamanaka, Y., Kuzuya, T. (2020). Ligand-induced luminescence transformation in AgInS2 nanoparticles: from defect emission to band-edge emission. J. Phys. Chem. Lett., 11, 3969–3974.

[2] Santara, B., Giri, P. K., Imakita, K., Fujii, M. (2013). Evidence for Ti interstitial induced extended visible absorption and near infrared photoluminescence from undoped TiO_2 nanoribbons: an in situ photoluminescence study. J. Phys. Chem. C, 117, 23402–23411.

[3] Swartz, C. H., Paul, S., Mansfield, L. M., Holtz, M. W. (2019). Absolute photoluminescence intensity in thin-film solar cells. J. Appl. Phys., 125, 053103.

[4] Toginho Filho, D. O., Dias, I. F. L., Duarte, J. L., Laureto, E., Harmand, J. C. (2005). Photoluminescence properties of Te doped AlGaAsSb alloys. Braz. J. Phys., 35(4), 1678–4448.

[5] Voguel, D. J., Kilin, D. S. (2015). First-principles treatment of photoluminescence in semiconductors. J. Phys. Chem. C, 119, 27954–27964.

[6] Hurtado-Castañeda, D. M., Herrera-Perez, J. L., Arias-Cerón, J. S., Reyes-Betanzo, C., Rodriguez-Fragoso, P., Mendoza-Alvarez, J. G. (2014). Aqueous sulfur passivation of N-type GaSb substrates studied by photoluminescence spectroscopy. Nat. Sci., 6, 963–967.

[7] Zehani, F., Sebai, M., Zaioune, H. (2012). Elaboration and characterization of AuCl semiconductor nanocrystals embedded in KCl single crystals. J. Mod. Phys., 3, 534–537.

[8] García, M. C., Mejia-García, C., Contreras-Puente, G. S., Hernández, J. A., López, J. L. (2000). Efectos de la variación de la energía de excitación en los espectros de fotoluminiscencia de películas de CdTe. Revista Mexicana de Física, 46(3), 265–268.

[9] Liang, H. F., Smith, C. T. G., Mills, C. A., Silva, S. R. P. (2015). The band structure of graphene oxide examined using photoluminescence spectroscopy. J. Mater. Chem. C, 3, 12484.

[10] Liqiang, J., Yichun, Q., Baiqi, W., Shudan, L., Baojiang, J., Libin, Y., Wei, F., Honggang, F., Jiazhong, S. (2006). Review of photoluminescence performance of nano-sized semiconductor materials and its relationships with photocatalytic activity. Sol. Energy Mat. Sol. Cells, 90, 1773–1787.

[11] Liang, H. F., Smith, C. T. G., Mills, C. A., Silva, S. R. P. (2015). The band structure of graphene oxide examined using photoluminescence spectroscopy. J. Mater. Chem. C, 3(48), 12484–12491.

[12] Peng, L., Wang, X., Coropceanu, I., Martinson, A. B., Wang, H., Talapin, D. V., Ma., X. (2019). Titanium nitride modified photoluminescence from single semiconductor nanoplatelets. Adv. Funct. Mater., 30(4), 1904179.

[13] Pawlak, M. (2019). Photothermal, photocarrier, and photoluminescence phenomena in semiconductors studied using spectrally resolved modulated infrared radiometry: physics and applications. J. Appl. Phys., 126, 150902.

[14] Samah, M., Khelfane, H., Bouguerra, M., Chergui, A., Belkhir, M. A., Mahtout, S. (2004). Optical responses of alkali-halide matrix (NaCl)-doped silver. Physica E, 23, 217–220.

[15] Yakota, M., Ito, A., Doki, N. (2016). Multi-color photoluminescence of terpyridine crystals. Adv. Chem. Eng. Sci., 6, 87–92.

[16] Walas, M., Lewandowski, T., Synak, A., M. Ł.,, Wojciech Sadowski, B. K. (2017). Eu3+ doped tellurite glass ceramics containing SrF2 nanocrystals: Preparation, structure and luminescence properties. J. Alloys Compd., 696, 619–626.

[17] Peng, W. Q., Cong, G. W., Qu, S. C., Wang, Z. G. (2006). Synthesis and photoluminescence of ZnS: Cu nanoparticles. Opt. Mater. (Amst.), 29(2–3), 313–317.

[18] Castro, R. C., Ribeiro, D. S. M., Páscoa, R. N. M. J., Soares, J. X., Mazivila, S. J., Santos, J. L. M. (2020). Dual-emission CdTe/AgInS$_2$ photoluminescence probe coupled to neural network data processing for the simultaneous determination of folic acid and Iron (II). Anal. Chim. Acta, 1114, 29–41.

[19] Rivera, W., Gálvez, C., Velasco, X., Morales-Acevedo. (2011). Control De La Fotoluminiscencia De Silicio Poroso Por Oxidación Anódica En Solución De C_2H_5OH/HF. Revista Colombiana de Física, 43(2), 369–372.

[20] Rojas Olmedo, G., Silva Gonzalez, R., Gracia-Jiménez, J. M. (1991). Influence of the thermal annealing and stoichiometry on the photoluminescence of cadmium. Revista Mexicana de Física, 37(2), 284–293.

[21] Swartz, C. H., Paul, S., Mansfield, L. M., Holtz, M. W. (2019). Absolute photoluminescence intensity in thin film solar cells. J. Appl. Phys., 125(5), 053103.

[22] Kameyama, T., Yamauchi, H., Yamamoto, T., Mizumaki, T., Yukawa, H., Yamamoto, M., Ikeda, S., Uematsu, T., Baba, Y., Kuwabata, S., Torimoto, T. (2020). Tailored photoluminescence properties of Ag(InGa)Se$_2$ quantum dots for near-infrared in vivo imaging. ACS Appl. Nano Mater., 3, 3275–3287.

[23] Peng, W. Q., Cong, G. W., Qu, S. C., Wuang, Z. G. (2006). Synthesis and photoluminescence of ZnS:Cu nanoparticles. Opt. Mater. (Amst.), 29, 313–317.

[24] Xiang, X., Zu, X. T., Zhu, S., Wang, L. M. (2008). Photoluminescence of SnO2 nanoparticles embedded in Al_2O_3. J. Phys. D: Appl. Phys., 41(22), 225102.

[25] Zehani, F., Sebais, M., Zaioune, H. (2012). Elaboration and characterization of AuCl Semiconductor nanocrystals embedded in KCl single crystals. J. Mod. Phys., 3(07), 534.

[26] Puruswottam Aryal, Abdel-Rahman Ibdah, Puja Pradhan, Dinesh Attygalle, Prakash Koirala, Nikolas J. Podraza, Sylvain Marsillac, Robert W. Collins, Jian Li. (2016). Parameterized complex dielectric functions of $CuIn_{1-x}Ga_xSe_2$: applications in optical characterization of compositional non-uniformities and depth profiles in materials and solar cells. Progress in Photovoltaics; Research and Applications. 24(9).

G. V. S. Subbaroy Sarma, Murthy Chavali, Periasamy Palanisamy, I. B. Shameem Banu

Chapter 7
Synthesis and Luminescent Applications of Rare-Earth-Doped Zinc Nanomaterials

Abstract: Some of the luminescent substances were focused on rare earth, especially the ion-doped lanthanide as well as transition-metal ion-doped hosting system. Through different kinds, like indistinct, metallic and even narrow films, several kinds with rare-earth-doped including rare-earth-free luminescent substances were created. Under many situations, certain characteristics with luminescence depend on its form with dopant employed with their activating agent. All elements of its rare earth provide some special yet critical influence over the human lifestyle. This unfulfilled 4f electron configuration of its rare-earth components allows itself to be included by generating numerous modern substances with different uses, like phosphors, magnetic substances, energy-storage components, including catalytic, according to different characteristics through luminosity, electric and magnetic fields, and telecommunications. Phosphors are mostly synthesized by chemical methods; these luminescent substances were examined for attaining creative yet alert substances including stable reaction, dry chemical, deposition and co-precipitation, oxidation and basic chemical path in several researches. A good proportion with luminescent structures depends on rare-earth particles including rare-earth-based substrate layers (that were being developed over the past couple of centuries). Most of such products have worked their path into applications. Upon utilization, rare-earth fluorescence can, across multiple situations, greatly enhance the efficiency of the products that they were used with.

Luminescent substances were widely utilized in solid-state illumination and screen technologies. Solid-state illumination development will improve effectiveness but rather reduces the output of glow diodes. Luminescent compounds would also be employed by detectors for maximum temperatures, but strain measurement, which is because of their ultra-sensitivity, is essential to determine several techniques. Luminescent substances had been extensively studied across biological applications, including glowing biological substances as well as even semiconductor nanomaterials, resulting in low hazard, strong physical and mechanical flexibility, narrow illumination patterns, high luminosity spectral levels and longish luminescence decaying rates. Rare-earth-based nanoparticles show significant benefits among most usable faintly glowing nanostructures. In this chapter, we will discuss synthesis and luminescent applications of rare-earth-doped zinc nanomaterials as "luminescent materials" involving such a broad variety of disciplines connected with rare-earth metals, nanoparticles, nanostructure, compositions with crystal membranes, nanosensors for

https://doi.org/10.1515/9783110676457-007

recording temperature, chemical sensors, biosensors, anti-counterfeit applications, security devices and so on.

Keywords: nanomaterials, nanophosphors, rare earths, luminescence, synthesis, applications

7.1 Introduction

Luminescence is "cold light," from different resources of electricity that may obtain region by ordinary as well as decrease temperatures. In this phenomenon, electrons presented in the ground state are excited to a higher state by absorbing energy, and then electrons return to the ground state by emitting energy like light. Luminescent fabric is also referred to as "phosphor." The name "phosphor" starts from the Greek language and signifies "light transport," to depict light-exuding or luminous materials; a phosphor is radiant, and emits electricity from an agitated electron as light. The agitation is because of the inclusion of electricity from an outside supply including some other electron, a photon. Excited electrons occupy a quantum position whose power is more than the minimal power of the ground situation. In semiconductors and insulators, the virtual lower state is regularly referring to electrons inside the valence band (VB), which is loaded with those electrons. The agitated quantum state is frequently located within the conduction band (CB) that is vacant and is divided as of the VB by way of an electricity hole referred to as the band gap, ΔE_g.

Phosphor is a strong faintly glowing duster that releases radiation by imposing certain exterior force like particle beams and sometimes UV radiation. Phosphor includes a number of lattices and is frequently called an "activator." It absorbs interesting radiations and rises to a higher state. In a few resources, the activator no longer absorbs the excitation radiation; however, different ions can additionally take in the exciting radiation and sooner or later transport themselves through that activator. Each hosting lattice passes its vibrational energy toward its activating agent within different objects, and therefore hosting lattice serves like a "sensitizer." Any form of oxide chosen among sulfide, aluminate, chromite, phosphate, halo phosphate and so on was indeed its hosting lattice.

In Germany, between the past nineteenth and mid-twentieth century, Philip E.A. Lenard and contemporaries worked on phosphors. They selected exclusive rare-earth ions as luminous ions in one-of-a-kind host materials [44]. We can have a look at the luminescence occurrence inside an environment resembling glowworms, mosquitos and even healthy marine microbes including the essence of the cavernous seas inside.

It is utilized in various fields through exceptional researchers everywhere on the planet, which incorporates pale history, geography, biomedical designing, science, material science and various creative applications for predominance control, examination and improvement.

7.1.1 Luminescent Centers

An extensive assortment of focuses conveys ascend to glow in semiconductors and protective substances, which includes rare-earth particles, transition-metal ions, exciton, pairs of donor–acceptors and ions with a state of d^{10} or s^2 electronic configuration. Similarly, pure molecules undergo glow on distinctive instances, rapid fluorescence aligned with singular-to-singlet spin-allowed progress and reasonable illumination from quadruplet-to-singlet developments. On the side of the rare earth, the spectra contain pointed strains that move up and down from simple electronic transitions terms, and the influence of the environments is felt specifically by their outcomes in the states of lifetimes. Hereafter in the ancient past to iridescence, its miles just in any case conversion of rare-earth glows, for example in which the impact of vibrations may be first ignored.

7.1.2 Phosphors

The substances both synthesized or within the natural shape showing iridescence of any kind is called phosphors. One example of ways a phosphor radiates light is that a darkish circle signifies an activator particle, encircled by a number of cross-sections.

7.1.3 Doping

The band hole energies of hardly any materials have emanation in data transmission that relates to the scene range. Be that as it may, substances with an incredibly extensive band opening might be made to discharge radiance inside the scene. It is feasible by utilizing the expansion of various iotas or defects into the gem. The extra particles, known as dopants, show an exceptional electron orbital structure when contrasted with the particle layer for the host. Additional energetic ranges are available within areas across each crystal throughout their doping fixing molecule. In the fabric's forbidden band hole, concentrations of energy will indeed coincide, which could cause electrons or gaps. The energy levels are closer across the CB; then, the dopants are known as the donors, and similarly, if they are near to the valance band, they are known as acceptors. Transitions among those ranges can provide upward thrust to noticeable iridescence and, in such occurrences, the dopants act as an activator. Each activator ranges from a single doped atom among 5,000 sample particles with around small when each dopant atom through 100,000 hosting atom through popular words.

7.1.4 Host/Self-Sensitized Luminescence

7.1.4.1 Sensitizer Sensitized Luminescence

Sensitizers additionally play a critical position within the luminescent substances and can take in the excitation strength and switch it to the activators. Each faintly glowing home then has been considerably advanced or larger. These luminescent materials are described as a kind of "host + sensitizer + activator." Each Ce^{3+} ion, for instance, is an excellent sensitizer against $LaPO4:Ce^{3+}$; Tb^{3+} phosphorus Tb^{3+} ions, which can assign every vibrational energy toward Tb^{3+}, also reveal inexperienced radiation.

7.1.4.2 Activator Luminescence

Maximum iridescent materials include latent hosts and brilliant particles. For instance, Y_2O_3 as a bunch fabric cannot luminescence without delay. While Eu_3+ particles were mixed through Y_2O_3, they will release red illumination. Then, we are defining the luminous materials with the "host + activator" type.

7.2 General Characteristics of Luminescence

The technique for radiance might be delineated in Figure 7.1.

Figure 7.1 depicts two kinds of going back to its initial state, they are radiative and non-radiative. This differentiation within iridescents had no limitation other than that which arises during thermal discharge due to electromagnetic radiation which has remain transformed into structural oscillations they either convey quality as warmth [6]. The productive glowing texture is one that throughout its thermal advances overwhelms each non-radiative models. Although the circumstance seems to be quite confused as portrayed in almost all of the iridescent substances (Figure 7.1), the exciting radiation is not constantly consumed by utilizing the activator however some other place.

7.2.1 Luminescence and Stokes Law

During its technique of glow, a portion of its power is ingested and re-transmitted as a light of an extended wavelength together as radiation is incident on a bit of material. A wavelength of light discharged in the arrangement of luminescence is a part of brilliant material and now not of the radiation incident.

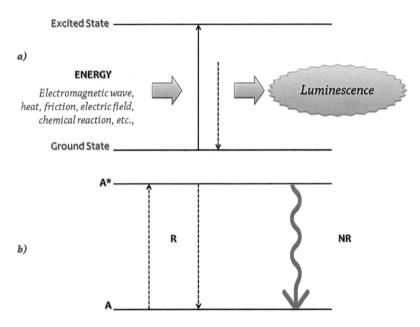

Figure 7.1: (a) Luminescence phenomenon and (b) energy steps of luminescent ion.

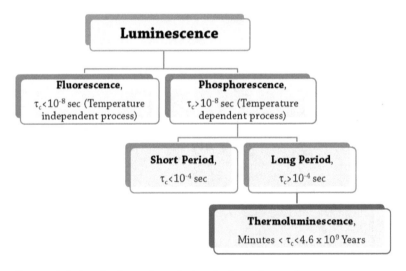

Figure 7.2: Order of radiance dependent on the term of emanation.

Based on the characteristics time (τ_c), luminescence is separated into two types. That is, if τ_c value is less than 10^{-8} s, they are named as fluorescence and then if it was greater than 10^{-8} s, as phosphorescence. Every system referred to has its importance and benefit within the discipline of technological know-how and generation [22]. Additionally, on the off chance that the put off duration was tons smaller, fluorescence and glow are much further notable but confusing. If the put-off time is tons shorter it is miles greater hard prominent difficult to separate among fluorescence and glow. In this manner, the glow is isolated into a brief period ($\tau_c < 10^{-4}$s) and extensive stretch (($\tau_c > 10^{-4}$s) phosphorescence. Fluorescence becomes temperature-free, even as rot with glow notable shows hearty temperature reliance. The own radiance wonders are demonstrated in Figure 7.2.

7.2.2 Method for Processing and Degradation of Exciting Systems

Methods with manufacturing as well as decomposition of vibrational levels may be depicted utilizing a vitality state chart called a Jablonski sketch. The Jablonski diagram (Figure 7.3) portrays the limit of the rest instruments for energized state particles.

Vibrational movements with particles to form a compound, the ground state S_0, lowest singlet S_1 and triplet states T_1 are made out of several vibrational conditions. Exactly when vitality tremendous than the HOMO-LUMO (most raised included sub-nuclear orbital, less least void sub-nuclear orbital) power contrast is incorporated straightforwardly into a particle, conveying both Inside of S1 regions and even greater singlet enhanced states S2 or S3, a larger vibrational region is transmitted. In a time scale of picoseconds, the stronger vibronic conditions of S1 decrease to the lower vibronic state of S_1.

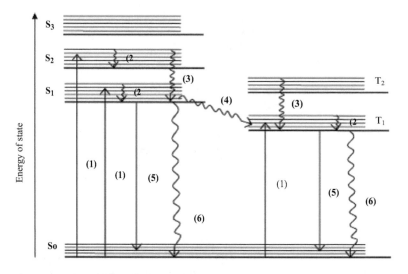

Figure 7.3: Jablonski diagram.

Through non-radiative internal change strategies, the more excellent singlet expresses that incorporate S_2 and S_3 pull again as much as the S_1 state. The states of the trio are usually rendered closer to S_1 to T_1 by intersystem crossing.

In the same manner, thermal developments embrace region as the modernized progression from the least permitted conditions of S_1 or T_1 to S_0 level. The convective movement from S_1 to S_0 is known to be a permitted transition, and in this way, the time estimation of the development will be in the request for certain nanoseconds. Then again, the time proportion of the T_1 to S_0 change is a dreadful part, more varying from miniature to milliseconds, because its technique was spin-forbidden. Therefore, a spread range seems like reflect photo including its atomic management range [16, 38, 41].

7.2.3 Design of Configurational Coordinate

The majority among iridescent substances involve an undeniable host crystal and an excellent particle known as an activator. It involves a little measure of polluting atoms that could be purposely conveyed into the host crystal. All the more frequently, with the help of the activator, the period exposure and vibrational capabilities of luminescent substances were calculated. Eu^{3+} and Eu^{2+} are defined by the configurational version of the coordinate.

Although the activator is essential for the radiance activities of phosphorus, optical properties are also influenced by the vibrations of the integrating molecules. To explain this fact, the configurational facilitate version is used to pick a glowing particle and its nearest neighbor destinations with a far off atom known as the radiant core. Thus, using a limited number or a sum of extraordinary proper directions, the enormously broad range of real vibrational methods of the system can be approximated, Q. It explains optical spots of a confined focus at the chance of capacity bends, which speaks to the general vitality of the particle in its lower or excited regions as an element of the structural directions. The entirety of the electron essentiality and the molecule imperativeness is completely here. The lower vibratory levels including energized regions were additionally foreseen for effortlessness is being equal. The model of interface coordinate (Figure 7.4) adaptation gives exceptionally valuable data consisting of:
a. Stokes' shift;
b. The thermal extinguishing of glow;
c. Temperature reliance and the widths of ingestion and emanation groups.

The parabolic state of twists is obtained from the shuddering movement of the bonds the majority of luminescent particle and nearest neighbor molecule which is believed to be symphonious and transmitted far imparted by Hooke's rule. The wide Q_0 induces the transfer of bigger Stokes and wider groups of absorption and emission. As proven

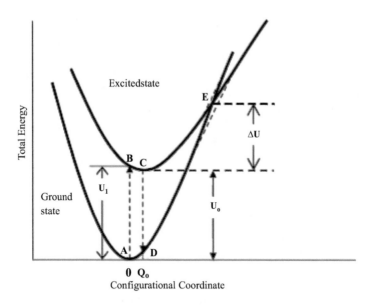

Figure 7.4: Configurational co-ordinate version.

in Figure 7.4, at the point, exactly any particle within the excited region will move non-radioactively toward its lowest energy via arrangement E across a warmer boundary once multiple parabolas join. Possibility of non-radiative advancement, W_{NR}, consistently relies upon the enactment energy expected to animate the electron from the energy-energized state's harmony position to the parabolic crossing point E:

$$W_{NR} = s. \exp(-\Delta U/kT) \tag{7.1}$$

Here k, s and ΔU represent Boltzmann's constant, frequency factor and thermal force, respectively.

The electron–hole transformation at optimal temperatures possibility is a better direct result of the exponential reliance on T. The electron–hole decay can be calculated even as the energy state's balance ability is positioned at the lowest state's external configuration organization twist. Under this situation, the energized state converges in unwinding from B to C, the ground condition. Even though W_{NR} is normally managed to utilize thermal relaxation approaches, it may be improved through the full power switch amongst glowing particles.

7.3 Luminescence Mechanism

7.3.1 Transferring Energy Throughout Its Down-Conversion Method

The glow of a visual substance is governed by the vitality movement system concerned among acceptor and contributor particles. Generally, resounding radiative quality switch, non-radiative power transmission and phonon helped vitality move between the donor (D) and acceptor (A), these are recognized force move strategies engaged with the lanthanide particles. They are represented in Figure 7.5(a, b).

7.3.2 Shift of Energies to Up-Conversion

A nonlinear anti-Stokes method where at least two vitality input siphoned photons is changed through radiation with highly energetic production [2]. These seem to be quite a couple extra noteworthy selective strategies concerned within the UT wonder. The additional typical energy switch strategies concerned with the UT procedure are defined.

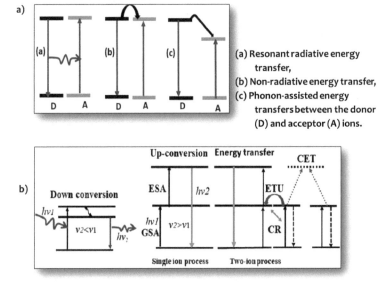

Figure 7.5: (a) Types of energy transfer and (b) down-conversion, single-ion process and two-ion process.

7.3.2.1 Grond-State Absorption

Even within the lowest energy, some rare-earth ions mendacity ingests the power of stimulation and elevated toward its greater energized level. Every single particle is used in such a technique.

7.3.2.2 Excited-State Absorption

Similarly, many atoms inside their stimulated category could have enhanced through vibration toward their better levels. On de-excitation, these particles throughout their highest level emit up-changed radiation in addition to their solitary particle measurement.

7.3.2.3 Up-Conversion of Energy Flow

Represented through the cooperation of various adjoining particles wherein one particle goes about as a sensitizer and retains the excitation vitality that is moved to the neighboring particle going about as an activator. The transmitted photon could have higher strength energized photons.

7.3.2.4 Cross-Rest Technique

On the off chance that if two sets of electricity tiers are having the same strength difference among them, there is a sure chance of electricity transfer and trade happening between the degrees beneath attention.

7.3.2.5 Co-Operative Power Transfer

At the point where two main particles have a link with each other within any metastable moderate degree and energize one particle toward its higher laying stage while then remaining particle gets de-energized, quite immoderate doping focus is required on such method.

7.4 Types of Luminescence

The different glow marvels are given names dependent on the sort of radiation used to energize the emanation (Table 7.1). Every framework expressed has its centrality and increase inside the field of science and technology.

Table 7.1: Luminescence wonder and strategies for excitation.

Luminescence phenomena	Methods of excitation
Photoluminescence	Light or photons
Bioluminescence	Bio-concoction strength
Sonoluminescence	Sound waves
Electroluminescence	Electric field
Chemiluminescence	Chemical vitality
Triboluminescence	Mechanical energy
Cathodoluminescence	Electrons or cathode rays
Radioluminescence	Nuclear radiation or ionizing radiation
Thermoluminescence	Ionizing radiation, UV and visible light

7.4.1 Photoluminescence

Photoluminescence has a very wide application for large-scale displays by whitening materials within cleaning detergent through laser video display systems. There is an exceptional kind of radiance that has a moderate rot with the emanation proceeding for quite a long time or hours. This type of glow is alluded to as enduring or constant iridescence and it is typically used in street insurance and goes out checking. Intrinsic or extrinsic luminescence may be labeled as photoluminescence [12, 42].

7.4.1.1 Intrinsic Luminescence

Inherent glow refers to a case, as the name suggests, where the iridescence originates from inside unadulterated crystals. This could be assembled among three groups:
a. Phosphor band-to-band: According to its recombination by each electron throughout any CB into a gap inside the VB, a pattern of luminosity occurs, conveying some band-to-band change. A certain type with excellent methodology has to be settled in every brand name substances at magnificently excessive temperatures;

under room temperatures, it is converted through exciton glowing. Occurrences with these substances were silicon, germanium including a couple of IIIb–Vb blends far-reaching of GaAs.

b. Cross-luminescence: When an electron recombines with an opening made in the most center band, cross-luminosity is provided. In antacid, soluble earth halides and double halides, this kind of iridescence is usually calculated.

c. Exciton luminescence: Every charged particle was a stable electron–hole set where a hole is bound by an energized electron. From the crystal, as it passes, it carries a couple of intensities, and the electron and empty recombine to give a glow. There are varieties of excitons. In which, the Wannier exciton, due to Coulomb interaction, usually inorganic semiconductors IIIb-Vb as well as IIb-VIb overcome each electron throughout its CB bond toward its hole within each VB. Even though its range of the elements of the electron and hole wave is steadier than the grid and it tends to be found in normal atomic gems comprising of anthracene, inorganic convoluted salts alongside tungstates and in uranyl salts.

7.4.1.2 Extraneous Iridescence

Extraneous iridescence alludes to glow caused by the introduction of impurities or defects in phosphorus and ionic material; this can be unallocated or reduced. This was unlocalized, whereas its available number of electron and accessible gaps within the hosting grid's VB also contains luminescence emissions. Conversely, when the excitation and outflow strategy for the iridescence is limited within a restricted iridescent core, the confined form occurs.

Luminescence is the object that absorbs optical radiation. This wonder is that incandescence, which is the discharge of radiation from material from it, is prevalent at a high-temperature system of excellence. Glow can emerge in an assortment of ways relying on a few unmistakable occurrences. In this manner, polymeric materials, synthetic, natural atoms, regular crystalline, even indistinct materials discharge radiance during sufficient situations.

Electrical vitality is changed over into radiation with a power of roughly 80%, yet the obvious light produced is substantially less than 10% of the absolute radiation delivered. Extreme radiation is especially as infra-purple vicinity or temperature. The range of radiation produced by a hot wire, or by some other part, is not relevant to the qualities of the material. All warm particles emanate mellow and warmth with fundamentally the same as properly defined with the aid of fashions based on a well-known blackbody.

Glow happens while cloth ingests radiation that achieves the progress of e^- from the VB to CB [8]. This is trailed by utilizing the de-excitation of e^- to get back through the VB using the radiance community that converts with electromagnetic radiation despite its unity [35]. The luminescence core can be either due to an inherent defect [40] or extrinsic defects [26].

Figure 7.8 recommends a schematic outline representing both intrinsic and extrinsic imperfections. Inborn deformities arise because a particle is transferred by its precise location with some substitution place within its lattice, which itself produces some illness factor, creating another flaw at the back of an opening imperfection [34]. Positive deformities occupy the donor phase, D (Figure 7.6), and the helpless imperfections involve the acceptor stage, A (Figure 7.6), inside the bandgap of a texture [7].

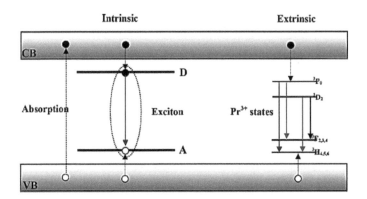

Figure 7.6: Emanation of both inherent and outward imperfections.

7.4.2 Fluorescence

Fluorescence seems to be their temperate discharge, to its restriction period $\tau_c < 10^{-8}$ s, where its surge occurs through any re-energized singlet phase whereas that flash ($\tau_c > 10^{-8}$ s) by another empowered trio condition where this spreading exists. To make a strong distinction between fluorescence and phosphorescence, its impact with temperatures on luminescence decline is to be studied. Photoluminescence is moreover applied to such a sparkle that is vivified by the light of concealing [6].

Another fluorescence illustration is in the latest devices for the development of clinical X-ray pics; a screen that creates a ton of seen mellow known as fluorescence when illuminated with X-beams is utilized to frame a picture that may then be shot with films that can be touchy to obvious light. This technique is extra touchy than the use of picture right now for a measure that X-rays, decreasing each dosage with X-rays toward any infected person as a result.

7.4.3 Phosphorescence

Electrons excited from the original radiation will make the effort in certain lower energy levels, they continued by decaying in a few materials. The rot can take such long time as scarcely any hours to hardly any days. Such fluorescence is alluded to brightness and the material keeps up to transmit seen light for quite a while after the first radiation has been turned off. On the off chance that the length is brief, around 10^{-4} s, at that point, the texture is brisk tolerance phosphor. Just by second, everything ends, an extended phosphorous lifetime is now and then known as luminous. Often objects displaying light are expected to glow. Many luminescent dolls, labels, signs in stores and aircraft watch knobs are coated for long stays.

7.4.4 Cathodoluminescence

The light discharge is an aftereffect of electron bar excitation. For this reason, an electron gun is used along with a cathode ray tube, TV besides an electron filtering magnifying instrument. Cathodoluminescence is caused by moderate pollution when being bombarded by intense irradiation of electron beams and it is the opposite of the photoelectric effect whereby the use of photon irradiation causes electron emission. Toward the start of the remainder of the century, the imperceptible cathode beams made by electron escapes through emptied pipes, developing illumination once they hit cylinder's crystal allotments. Electrons are the cutting edge name for cathode beams, and the cathodoluminescence was held by this kind of radiance. This is one of the beneficial advantages of iridescence. The electron-magnifying instrument utilizes light emissions to gracefully high choice pictures of modest example. On a couple of occasions, cathodoluminescence from the case is created by the beam. It was particularly valuable for the investigation of stone particles everyplace; the existence of sign proportions of progress metallic parts will make the mineral offer smooth incredible covering. Usually, the reality of the clue detail cannot be recognized from every other perspective. Likewise, the ever-present video shows that tube utilizes a light emission to specifically energize purple, green or blue phosphors to show concealing reviews.

Cathodoluminescence occurs because the deformation on a semiconductor of excessive electron beams may lead to electrons migrating through its zone of valence across their CB, giving up a gap behind. Once every electron, as well as a vacancy, melds properly, a photon can be transmitted. The potency of photon and the chance of emanating a photon and not a phonon rely upon the surface, its perfection and every level of flaw. Within such condition, every "semiconductor" breakdown should become a general rule, and be almost any non-metallic fabric. Regarding band structure, old-style semiconductors, covers, earthenware production, gemstones, minerals and glasses might be dealt with in an indistinguishable way. As of late, cathodoluminescence executed in electron magnifying instruments is getting utilized for contemplating surface plasmon reverberation in metallic nanoparticles [11, 45].

7.4.5 Chemiluminescence

Chemiluminescence is the emission of light through the discharge of electricity from a chemical reaction. The outflow of light utilizing the release of intensity from a synthetic response is alluded to as chemiluminescence. The instrument for such outflow is the change of mixture power, delivered in an unimaginably exothermic response, into light quality inside the noticeable region. In a couple of compound responses, power might be shifted with e^- inside its chains of materials. This terminates in moderate discharge where some e^- fall downward which pull back energized systems. A little bit of this response continues slowly so that the light can be transmitted for an infinite period. This is indicated in the form of chemiluminescence.

Chemiluminescence is shown through an assortment of life forms and the substance response, for the most part, includes oxidation. This sort of gentle producing substance is known as luciferin. This is a natural particle with two hydrogen molecules connected with the essential picture as LH_2. In an aligned position, luciferin is metabolized into L=O with the significant asset of the chemical liable for the cell quality, adenosine triphosphate and an exceptional impetus protein. Every observable illumination particle was emitted precisely since that shifts through its lowest energy. The gentle is regularly light blue in shading, even though diverse substance conditions can shift the concealing. This light-making methodology is expected to advance as a small aspect of standard oxidation–reduction reactions that separate energy from metabolites. A few technological particles comprehensive of luminol and cyalume are the foundation of efficiently to be had chemo brilliant items. Normally, this light-making methodology progressed as a little aspect of the standard oxidation-decrease reactions those different forces from supplements.

7.4.6 Electroluminescence

Electroluminescence is the powerful technology of mild in an applied electric-driven region or plasmas in a non-metallic stable or fuel. Another type of electro-glow is delivered through certain crystals while a cutting-edge electric front line through them. In such a case, the flow of energized electrons that possess power levels worried about synthetic bonds inside the crystal. This effect is called electroluminescence as the energized electrons rot again to their ground state when they transmit light.

Few amazing electroluminescence systems come from a crystal. Around 40 years ago, it was such a gentle idea that for certain home applications ordinary lights could be exchanged as electroluminescent coatings could be added to blocks, roofs, even blinds went to the hues scale. Unfortunately, it was not possible to resolve many sensitive problems, which includes performance, and that high-frequency AC becomes important to excite the material. However, along with the mobile shows other than light-emitting diodes (LEDs), the mild emitting diode, which works on an extraordinary principle, has now become a commonly used method of electroluminescence.

7.4.7 Ionoluminescence

The visible mellow produced when fast particles slam through natural and inorganic mixes is a more prominent, specific method of assembled radiance. This phenomenon is referred to as ionoluminescence. Early use of ionoluminescence converted into glowing dials of the clock relied on a dangerous choice to make light that required radioactive compounds. Along with radium, radioactive fabrics were mixed with such a substance displaying luminescence along with zinc sulfide. It released α-debris and other radiation as the radium decays. This energizes electrons to give light inside the luminescent material, which is entirely available because the gentle perseveres uncertainly, restrained simplest via the 1/2 availability with an isotope containing radium utilized, 226 Ra, that is, 1,600 years. At a brilliant dial depict fabricating office, because of licking their paintbrushes to get an incredible brush point, the laborers were exposed to radiation. This releases "α" trash and other radiations as the radium decays. At an iridescent dial depict fabricating office, because of licking their paintbrushes to get an amazing brushing level; intrinsic radioactive emission was exposed to several laborers.

7.4.8 Lyoluminescence

Each emission by illumination is indicated through lyoluminescence while dissolving a strong into a liquid dissolvable. This lyoluminescent effect is seen at the stage when solid samples that have been strongly lit up via ionizing energy were broken down within the water. The accumulated amount of light produced within its substance increases toward some precise point termed an immersion estimate relative to the overall emission portion obtained throughout each substance. Lyoluminescence is recognized by various γ-light substances; these include flavors, compact dairy, sauces, wool and even paper. By disintegrating solids in a solution that typically contains chemoluminescent mixes, such as luminol, the lyoluminescence capacity can be increased. Thus, such mixes are called luminescence sensitizers [33].

7.4.9 Radioluminescence

At the point when a propelling molecule of radiation slams into a particle or atom, radioluminescence happens, enhancing the angular electrons toward the prevailing intensity level. At that point, the electron re-visits its degree of ground vitality by generating more quality like photons by radiation. It is often delivered with an ionizing radioactivity guide. Many materials possess biological compounds such as X-rays, γ-rays or cosmic rays, which release mild once bear with emission they turn as sensors for excessive radiation of power. Radioactive factors can emit α-particles, γ-rays and electrons.

7.4.10 Sonoluminescence

It seems to be the energy released through acoustic pressure and due to stimulation, produce ultrasonic vibrations. Rayleigh suggested that the explanation for this was the collapse of air bubbles [27]. This phenomenon has been discovered by the University of Cologne as a result of sonar paintings. In a photographic developer fluid tank, H. Frenzel and H. Schultes mounted ultrasonic transducers that accelerate the developed procedure. Instead, upon growing, they observed splotches onto their film but realized, for their ultrasound turned enabled, their pores inside its liquid emitted illumination. Because with this complex world for such a large range in low droplets, it has become too difficult in early experiments to examine the effect. Such an effect has been called multi-bubble sonoluminescence.

It emerges while an acoustic flood with enough power initiates a vaporous space inside a fluid to disintegrate expedient. This cavity can likewise take the state of a prior air pocket or might be created utilizing a technique called cavitation. Sonoluminescence inside the lab might be made to be solid, with the goal that a solitary air pocket will enhance and self-destruct on numerous occasions in an intermittent style, radiating an explosion of mellow every time its breakdown. A standing acoustic wave is an installation in a liquid for this to occur, and the bubble will stay at a status wave strain anti-node. Resonance frequencies rely upon their form and length of the bubble-included region [10].

7.4.11 Bioluminescence

Bioluminescence is otherwise called "staying mellow" and the deep seas may be the most gigantic instances of this marvel that might be situated in the remote oceans. In the murkiness of the seas wherein, sunlight cannot accomplish, a few dwelling creatures produce light as the consequence of substance responses.

7.4.12 Triboluminescence

In this sort of luminescence, mechanical power generates light. The breaking of chemical bonds within the material causing mild contamination, including in sugar or silicon crystals, may result in pulling back, tearing, scratching, grinding or rubbing. It is likewise named mechanoluminescence.

7.4.13 Thermoluminescence

Thermoluminescence is often named thermally stimulate luminescence. It is the iridescence that is thermally enacted by an alternative technique like α, β, γ, UV and X-beams after start illumination. It is not, for the most part, to be stressed over hot energy: The handiest warm stimulation activates its emergence with harmony through some wellspring with vibrations presented mostly for substance.

7.4.14 Radioluminescence

It is a miracle where the only illumination was transmitted to any surface using ionizing energy barrages, such as β-rays, X-ray beams and including radioluminescence referred γ-beams.

7.5 Synthesis of Luminescence Materials

Numerous methods have been used in recent years to simplify the synthesis of phosphorus. These can be alluded to widely as "new synthesis." The tale blends are not the best simpler techniques for phosphor synthesis. Numerous epic procedures were directed for the amalgamation of those phosphors which incorporate; stable state diffusion, solvothermal, dry concoction, precipitation and co-precipitation, combustion, clear route of a mixture, a technique of sol–gel, production of molten salt and many others.

7.5.1 Method for the Diffusion of Solid States

In this technique, the parts were allowed that respond through its diffusion process. With some kind of fair reaction, response duration and heat perseveres. This probably would not be reasonable at whatever point to bring down the last adequately.

E. G. numerous aluminates cannot be fashioned even around ranges under 1,400 °C, solid-phase propagation. Typical diffusion of the solid-state is in Figure 7.7.

Initiating hosting with concentration >98% being employed with source substances during the production containing certain phosphors. The reagents were said something stoichiometric extents; completely ground, muddled together an agate mortar, and calcined at a 100 °C for 5 h in a secured alumina pot. These calcined substances have been re-ground inside the mortar and re-terminated at 100 °C for 5 h under the reductive feeling of 5% H_2/95% N_2 to lessen 1+ particle to 2+ particles individually [46].

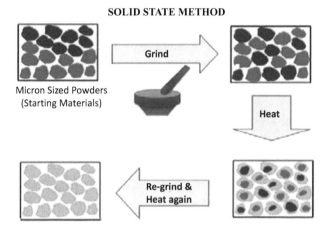

Figure 7.7: Solid-state diffusion method.

7.5.2 Solvothermal Synthesis

The solvothermal approach can be characterized as concoction responses or within elevated temperatures, changes in a solvent. Under such circumstances, certain basic physico-concoction features with solvents will specifically enhance dynamic dispersion among material organisms in a unique way. Such a solution should be able to dissolve its materials combined in this manner securing the reactants in the fluid area without the utilization of a dissolvable. There might be some issues with this framework being practiced. The softening purposes of the constituents like oxide are generally exceptionally high.

Solvothermal combination is a technique for preparing the dissemination of radiant materials that incorporate semiconductors, earthenware production, metals and polymers. As shown in Figure 7.8, it is similar to the hydrothermal method. The best ability is that of the antecedent arrangement which probably would not be commonly fluid. The technique includes the utilization of a dissolvable underneath moderate to high strain and temperature that empowers the interchange of forerunners all through the blend.

If water is utilized, the methodology is alluded to as an "aqueous blend." The cycle might be utilized to set up an incredible arrangement, calculation alongside skinny films, large-scale chemicals, individual including nanopowder clusters. This technique will also be used to plan stable crystalline and metastable conditions of specific substances which cannot even be formed by certain artificial ways without issues. Most of the literature about solvothermal processing had centered on nanomaterials throughout its end of the century such assessment will therefore spotlight a few developments in the nanocomposites, solvothermal blend [13].

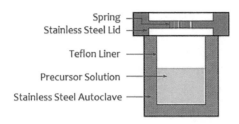

Figure 7.8: Solvothermal synthesis setup.

Utilizing this approach benefits their advantages with each the sol–gel [24] besides aqueous courses [1]. It considers the appropriate control over the size, structure dispersion and crystallinity of metallic oxide nanoparticles or nanostructures. While using changes through some test parameters, such characteristics are being changed, comprising of reaction temperature, response time, dissolvable kind, surfactant type, and antecedent sort. The nanostructured titanium dioxide, graphene, carbon, chalcogenides, and other materials were developed in the laboratory using solvothermal synthesis.

7.5.3 Molten Salt Synthesis of Compounds

Molten salt synthesis (MSS) is a demonstrated elective way to blend an immense assortment of mixes [36]. MSS of mixes includes blending reactant by an overabundance of salt, in a cauldron, underneath a changing climate or beneath the air. The containers are produced using alumina (Al_2O_3), zirconia (ZrO_2) and SiO_2, to prevent certain relation through their container substances, depending on the decision including its liquid salt. After that container was heated over its softening element within the salts in a furnace, reactions break up but also react through a reaction. That salt serves mostly like a dissolvable reaction through that step, and the liquid salts would be frozen then cleaned using H2O, liquor or physically separated for obtaining that final absolute product [15].

7.5.4 Wet Chemical Method

This mixture of moist concoction being the most optimal method is utilized throughout its blended state as best, synthetically accurate, yet unadulterated, singular pigments with synthesized conditions. Instead of their ability to generate objects at 800 °C, such technique was seductive. Its semi, as well as artificial reaction movements, may continue quickly in the fluid state. One can sum up the favors of the wet substance amalgamation are ease combination approach, as the low temperature is needed, power-sparing strategy on account of low-temperature prerequisite and the comparison to their solid-state dispersion process, burning technique and sol–gel approach, the best procedure toward any homogenization by component salts [4]. The clammy concoction method was used to prepare its standard multi-crystalline-joined composite phosphors, for example, component hosts over molar ratio proportions were broken up by profoundly refined deionized water again but dissipated until blend becomes anhydrous. That material phase of each substance phase was vaporized for 8 h at 80 °C. The resultant polycrystalline mass changed into a crucible to good flotsam and jetsam in a container. For additional portrayals, the powder is utilized.

7.5.5 Combustion Synthesis

The self-heat generating process is also known as combustion synthesis. With several cases of these unorthodox approaches, higher activation concentrations were carried out from collecting its reactivity mixture through excellent shape. In this system, the reactants are very well blended but were molded on any strip. Unique quit of block heats the use of any sort of flame to excessive temperature. When the exothermic response starts, it extends over the width of the block. In such reactions, concentrations are very large as 7,000 °C for the supply of nitride and carbide materials. Within their burning blend, the exothermic interactions among metallic nitrates including fuels along with urea were violated. Nitrate can be easily adapted to the proportion of fuel to alter its produced temperatures. Many remarkable concentrations were obtained once such proportions are modified with the ultimate objective of reacting and reducing valences of the oxidizers and gas.

It is also known as Thermal explosive synthesis. Al_2O_3 is used because of the aluminum effortlessly using a dissolving segment softening variable of 1,800 °C and the liquefying states of the greater part of different oxides are significantly higher. Indeed, even the strong state dispersion must be completed at temperatures appropriately over 1,200 °C and ordinarily over 1,700 °C for this reason. Aluminates can be processed by the synthesis of burning amalgamation involving metallic nitrate and urea sometimes without problems.

7.5.6 Simple Chemical Synthesis

The procedure implemented [3] through the conversion of natural ligands in specific is pointed to as simple chemical synthesis, which is a basic, low cost and periodic time-ingesting strategy. The required volume of a substance is diluted in alcohol and de-protonated by including a basic response to the NaOH solution. By dissolving the accompanying metallic salt over that nitrate in double purified H_2O, a metallic salt solution is prepared. Both the appropriate responses are mixed with energetic mixing. As fast as every arrangement is mixed, chelate accelerate appears to be inside the appropriate response. The acquired precipitate is filtered by using high-quality filter paper and then cleaned over many hours with very powerful light which drives out from their dampness. The reaction mixture of the substance was further separated but fixed for investigation throughout each container. In the fundamental zone, the pH is maintained around 10 within its response. The proportion of salts with anti-inflammatory pharmaceutical products was provided with 1:2 doses. The combined chelates can be sorted out via disintegrating their substituents in 99:1, 95:5, 90:10 and 80:20 proportions.

7.5.7 Precipitation and Co-precipitation

A few phosphors have been combined through sodden substance strategy helped by utilizing the precipitation approach as in Figure 7.9. In the double-distilled water, the stoichiometric host and dopant amounts are taken and carefully combined one by one and the mixture is well stirred on the magnetic stirrer to make the solution clear. In certain cases, the acquired material is cleaned with filtered water which breaks off their waste material and enabled them to dry on the cool, excellent filtered sheet for 24 h. It is possible to prepare both the pure and doped phosphorus samples to buy in such a manner. Completely test specimens appeared through fine powder; however, for measurements that produce pleasantly powdered, it took very small grinding [5].

Co-precipitation, at some stage in recrystallization and strong arrangement growth, was among several more extreme realistic strategies besides consolidating the following RE factors toward nanophosphors against any small sample conveyance. In contrast with various strategies, the co-precipitation technique does no longer require severe reaction conditions, expensive gadget, or complex methodology. Once in a while, glass-like nanophosphors might be gained promptly through co-precipitation excluding the calcination process or post-strengthening technique.

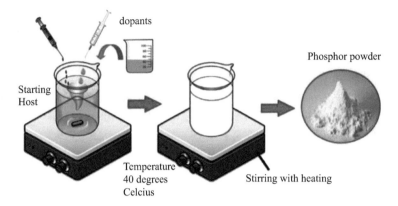

Figure 7.9: Synthesis of phosphor by precipitation technique.

7.5.8 Sol–Gel Method

This approach has been a crucial mechanism that is used now daily. The exceptional types of materials produced using this process are hyper or circular structured pigments, composite fiber, nanoporous synthetic molecules, homogenous ceramics including glasses or highly flexible nanotubes substances. The use of the sol–gel technique is used in four stages [14] for the synthesis of nanomaterials:
a. In addition to metallic alkoxides, the starting materials used within the sol path were usually inorganic substances or organic products of steel.
b. The compound is characterized by a set with hydroxylation but still polycondensation processes through its traditional sol–gel response toward creating a homogeneous dispersion labelled as a gel.
c. In this step, warmth treatment is provided to the gel emulsion that dried itself back toward its stage that is required with similar manufacturing. This solution gets completely mixed through a centrifuging machine to the chosen shape or size using a spin coating system with the aid of a suitable agent, washing is carried out several times before extracting its material.
d. We can produce the trash or films of the favored shape and scale using particular strategies. The centrifuge motion, for example, is used to obtain zirconia nanoparticles via spinning coated or spray-coated on a chunk of a substrate, skinny displays could be produced.

The sol–gel approach is dependent mostly on conversion with this sol portion produced through metal alkoxides or organometal precursors. Such a sol, reaction comprising suspended debris, has been polymerized to form a moist gel over extreme room concentration. The corresponding advance is that the dissolvable is wiped out by each liquid processing with proper warmth procedure [32].

7.6 Applications

There are several luminescent material sets, but we can better concentrate on a handful, which can be new technological advances. That is because of the continuous luminescence for home illumination, organic imaging luminescence, thermal sensors, glow electrodes transferred to white phosphorus and phosphor coating with any TV broadcast.

7.6.1 Home Lighting

Industrious iridescence creates its elective light which was financially savvy but also vitality traditionalist. This offers any chance to supply every lamp with no electrical connection that provides light. Such a bulb might be made from a constant radiant material, might be set out of entryways during the day to absorb the sunshine, after which situated returned only within the building, which that could keep within its non-appearance with the supply for stimulation that illuminates [23].

7.6.2 LED

LEDs are a type of solid-state illumination period centered using inorganic materials which move from solidarity toward illumination. The use of InGaN, which employs a crew of scientists headed by Isamu Akasaki, Hiroshi Amano and Shuji Nakamura [20], was recently completed with a blue-emitting light diode. An LED showing a good white emission (Figure 7.10) compared to the conventional incandescent lighting apparatuses converted into later performed with the help of transforming the blue LED into white LED [43] diode pairing to yellow phosphate radiation, $Y_3Al_5O_{12}:Ce^{3+}$ [39].

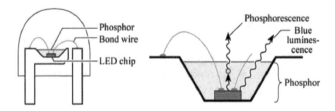

Figure 7.10: Phosphor-converted design with white LED.

7.6.3 Television Display

Each among imperative bundles with phosphors is throughout showcase innovation. They relied on TV for watching reports but also occasions remain, with complete shades. Both coloring and online videoconferencing use mobile devices to capture minutes. Every one of those is made feasible by utilizing phosphors for the show. There are a few advances that are utilized to harvest special types of TV programs, going from cathode beam tubes [25] to fluid precious stone show [21].

7.6.4 Photovoltaic Cells

Using the photovoltaic (PV) effect, a PV or solar mobile device turns daylight electricity into electricity without delay. While promising, less efficiency and high manufacturing costs afflict PV cells, which shun its usage on a broad level. There had also been many attempts which improve its performance of PV panels to transmit the rate of electricity. Luminescent solar concentrator could have been joined to win over their halfway utilization including its sun-powered range to develop the execution of silicon-based PV panels. Besides Si-based solar panels, almost all and perhaps dye-sensitized PV modules were recent topic research that could specifically interact against Si-based PV panels. However, like with Si–PV panels, those were never so efficient or even costly either [9]. The quest toward good PV panels was a nonstop framework. Fig suggests PV response of the content of TiO_2 and SEQD/TiO_2 hybrid movies within unique QDs content. The development of a new PV matrix is an opportunity for Si-based PV cells.

7.6.5 Luminescent Sensors

Glowing substances may be used by detectors for estimating unnecessary temperatures including strains that are difficult on varying degrees due to their scandalous affectability. The thermal dependence of Nd^{3+} in fluorotellurite glass was measured over any range of temperatures between 300 and 650 K through means of a test method of 885 and 815 nm discharge spectra [18]. Lower focuses demonstrate a superior affectability closer to the temperature changes. There is an interest in precise and solid strain scale for programming in exorbitant strain try. Laser-actuated ruby fluorescence R_1 lines are their more widely used stress estimation levels up until megabytes formed. Pinnacle shifts happen as an attribute of the weight which makes it fitting to utilize it as a pressure sensor.

7.6.6 Others

Removed articles can be estimated by utilizing optical temperature sensors. Non-contact humidity sensor detection has been identified with the use of RE phosphorous dopant particle due to their existence for heat linked vitality stage. In the calculation of the fluorescence intensity ratio (FIR) investigation, these TC levels were used. The FIR technique is reasonably adapted to each estimation with any temperatures throughout their RE sample do-pants. This strategy becomes additionally relevant with estimating physical warming toward laser-based hypertension applications. Up-conversion sensors fixed with RE, particles were ending up being improved applicants herein respect [37]. Photoluminescence planning empowers the following of medication conveyance to survey the viability of the medication discharge. Utilizing zinc gallate doped with Cr particles, develop its pathway for illuminated drugs via digestive storey, and forward with orally using [31]. Phosphor substances capture their energies once activated through light exposure like X-rays, which emit themselves through that direction with illumination.

An only phosphorous substance that gives its radioactivity exposure that longitudinal response may be suitable with the scintillation activities. The efficiency of the scintillator was calculated through its size of scattered electrons for every event of incoming light including the emissions of blue photons. Alumina has been mostly with the fascinating substances besides reflectance uses due to increased radioactive durability and a deep insight including its visible-light luminosity process [17, 19, 28–30]. With a very highly irradiating situation like an advanced nuclear reactor, alumina may be ideal for use as a scintillator.

7.7 Conclusion

Some quick synthesis procedures including implementation among solid-state phosphor substances addressed in this chapter. There are different strategies for planning glowing materials like solid-state diffusion, solvothermal, liquid salt, wet compounds, ignition, basic synthetic, precipitation, co-precipitation and sol–gel process that have researched for acquiring specialized but also smart substances. With luminescence applications typically restricted toward illumination, screens, scintillators even phosphor indicators, the potential for nanoparticle luminescence across different optoelectronics will be largely based on fracture synthesis, lower production costs including phosphor amplitude performance.

References

[1] Andersson, M., Österlund, L., Ljungström, S., Palmqvist, A. (2002). Preparation of nanosize anatase and rutile TiO_2 by hydrothermal treatment of microemulsions and their activity for photocatalytic wet oxidation of phenol. J. Phys. Chem., 106, 10674–10679.
[2] Auzel, F. (2004). Upconversion and anti-stokes processes with f and d ions in solids. Chem. Rev., 104, 139–174.
[3] Awad, S. H. (2012). Synthesis and characterization of some mixed–ligand complexes containing salicylic acid and pyridine with some metal ions. Al-Nahrain J. Sci., 15, 23–29.
[4] Bhake, A. M., Nair, G., Zade, G. D., Dhoble, S. J. (2016). Synthesis and characterization of novel $Na_{15}(SO_4)_5F_4Cl$: Ce^{3+} halosulfate phosphors. Lumin. J. Biol. Chem. Lumin., 31, 1468–1473.
[5] Bhake, A. M., Zade, G. D., Nair, G., Dhoble, S. J. (2015). Investigation of photoluminescence properties of reddish-orange emitting $Na_2ZnP_2O_7$: Eu^{3+} phosphor. Int. J. Lumin. Appl., 5, 224–228.
[6] Blasse, G., Grabmaeir, B. C. (Eds.). (1994). Luminescent Materials. Springer Verlag, Berlin.
[7] Breithaupt, J. (Eds.). (2000). New Understanding Physics for Advanced Level, 4th. Nelson Thornes Ltd, London.
[8] Busch, D. D., Nikon, D. (Eds.). (2006). 200 Digital Field Guide. Wiley Publishers, Indianapolis, IN.
[9] Correia, S. F. H., Bermudez, V. Z., Ribeiro, S. J. L., Paulo, S. A., Rute, A. S. F., Luís, D. C. (2014). Luminescent solar concentrators: challenges for lanthanide-based organic-inorganic hybrid materials. J. Mater. Chem. A, 2, 5580–5596.
[10] Didenko, Y. T., McNamara, W. B., Suslick, K. S. (2000). Effect of noble gases on sonoluminescence temperatures during multi-bubble cavitation. Phys. Rev. Lett., 84, 777–780.
[11] Garcia De Abajo, F. J. (2010). Optical excitations in electron microscopy. Rev. Mod. Phys., 82, 209–275.
[12] Garlick, G. F. J. (Eds.). (1949). Luminescence Materials. Clarendon Press, Oxford.
[13] Gersten, B. (2005). Solvothermal synthesis of nanoparticles. Chem-files, 5, 11–12.
[14] Kalyani, N. T., Sawde, S. M. (2015). Applied Physics-II, 1st. Himalaya Publishing House Pvt. Ltd., Nagpur, ISBN: 978-93-5202-404-9.
[15] Kimura, T. (2011). Chapter 4-Molten salt synthesis of ceramic powders. In: Costas, S. (Eds.). Advances in Ceramics-Synthesis and Characterization, Processing and Specific Applications. IntechOpen, Japan.
[16] Kitai, A. H. (Eds.). (1993). Solid State Luminescence. Chapman & Hall, London.
[17] Kortov, V. S., Pustovarov, V. A., Shtang, T. V. (2015). Radiation-induced transformations of luminescence centers in anion-defective alumina crystals under high-dose irradiations. Nucl Instrum Methods Phys Res B, 353, 42–45.
[18] Lalla, E. A., Leon-Luis, S. F., Monteseguro, V., Rodríguez, C. P., Cáceres, J. M., Lavín, V., Mendoza, U. R. R. (2015). Optical temperature sensor based on the Nd^{3+} infrared thermalized emissions in a fluoro-tellurite glass. J. Lumin., 166, 209–214.
[19] Lederer, S., Akhmadaliev, S., Von-borany, J., Gütlich, E., Lieberwirth, A., Zimmermann, J., Ensinger, W. (2015). High-temperature scintillation of alumina under 32 MeV $^{63}Cu^{5+}$ heavy-ion irradiation. Nucl Instrum Methods Phys Res B, 359, 161–166.
[20] Leon-Luis, S. F., Mendoza, U. R. R., Lalla, E., Lavin, V. (2011). Temperature sensor based on the Er^{3+} green upconverted emission in a fluoro-tellurite glass. Sens. Actuators B, 158, 208–213.
[21] Li, Z. J., Zhang, Y. J., Zhang, H. W., Fu, H. X. (2013). Long-lasting phosphorescence functionalization of mesoporous silica nanospheres by $CaTiO_3$: Pr^{3+} for drug delivery. Micro. Meso. Mater., 176, 48–54.

[22] Mckeever, S. W. S. (Eds.). (1985). Thermoluminescence of Solids. Cambridge University Press, Cambridge.
[23] Noto, L. L., (2015). Persistent luminescence mechanism of tantalite phosphors, PhD Thesis, University of the Free State, South Africa.
[24] Oliveira, M. M., Schnitzler, D. C., Zarbin, J. G. A. (2003). (Ti,Sn)O_2 Mixed Oxides Nanoparticles Obtained by the Sol–Gel Route. Chem. Mater., 15, 1903–1909.
[25] Pandey, A., Rai, V. K., Kumar, V., Kumar, V., Swart, H. C. (2015). Upconversion based temperature sensing ability of Er^{3+}–Yb^{3+}codoped $SrWO_4$: an optical heating phosphor. Sens. Actuators B: Chem., 209, 352–358.
[26] Pelant, I., Valenta, J. (Eds.). (2012). Luminescence Spectroscopy of Semiconductor. Oxford University Press, Oxford.
[27] Puttermann, S. J. (1995). Sonoluminescence: sound into Light. Sci. Am., 272, 32–37.
[28] Rahman, A. Z. M. S., Awata, T., Yamashita, N., Xu, Q., Atobe, K. (2009). Optical vibronic emission spectra for irradiation-induced F aggregate centers in single crystal alpha-Al_2O_3. Radiat. Eff. Defects Solids: Incorp. Plasma Sci. Plasma Technol., 164, 692–698.
[29] Rahman, A. Z. M. S., Awata, T., Yamashita, N., Xu, Q., Atobe, K. (2009). Optical vibronic spectra in reactor neutron-irradiated α- Al_2O_3. Phys. Proc., 2, 551–557.
[30] Rahman, A. Z. M. S., Kobayashi, T., Awata, T., Atobe, K. (2010). Thermoluminescence of α-Al_2O_3 by neutron irradiation at low temperature. Radiat. Eff. Defects Solids: Incorp. Plasma Sci. Plasma Technol., 165, 290–297.
[31] Rai, V. K. (2007). Temperature sensors and optical sensors. Appl. Phys. B, 88, 297–303.
[32] Rajaeiyan, A., Bagheri-Mohagheghi, M. M. (2013). Comparison of sol-gel and co-precipitation methods on the structural properties and phase transformation of γ and α-Al_2O_3 nanoparticles. Adv. Manufac., 1, 176–182.
[33] Reynolds, G. T. (1992). Lyoluminescence. J. Lumin., 54, 43–69.
[34] Ropp, R. C. (Eds.). (2012). Luminescence and the Solid State. Elsevier Science Publishers, Amsterdam, Netherlands.
[35] Salinas, C., Bracamonte, G. (2018). Design of Advanced Smart Ultra-luminescent Multifunctional Nanoplatforms for Biophotonics and Nanomedicine Applications. Front. Drug Chem. Clin., 1, 1–8.
[36] Sheikh, R. A., (2016). The synthesis of cementitious compounds in molten salts. PhD Thesis, Department of Chemical Engineering, University College London.
[37] Shen, Y., Wang, X., He, H., Lin, Y., Nan, C. W. (2012). Temperature sensing with fluorescence intensity ratio technique in epoxy-based nanocomposite filled with Er^{3+}-doped 7YSZ. Comp. Sci. Technol., 72, 1008–1011.
[38] Shionoya, A. S., Yen, W. M. (Eds.). (1999). Phosphor Handbook. CRC Press, Boca Raton, FL.
[39] Singh, S. K., Kumar, K., Rai, S. B. (2009). Er^{3+}/Yb^{3+} Codoped Gd_2O_3 Nano-Phosphor for Optical Thermometry. Sens. Actuators A, 149, 16–20.
[40] Stavale, F., Nilius, N., Freund, H. J. (2013). STM luminescence spectroscopy of intrinsic defects in ZnO (0001) thin films. J. Phys. Chem. Lett., 4, 3972–3976.
[41] Tsutsui, T. (1999). Fundamentals of luminescence. In: Shionoya, A. S., Yen, W. M. (Eds.). Phosphor Handbook. CRC Press, Boca Raton, FL.
[42] Vij, D. R. (Eds.). (1998). Luminescence of Solids. Plenum Press, New York, NY.
[43] Wade, S. A., Collins, S. F., Baxter, G. W. (2003). Fluorescence intensity ratio technique for optical fiber point temperature sensing. J. Appl. Phys., 94, 4743–4756.
[44] Wiedemann, E. (1888). Uber Fluorescenz und Phosphorescenz, I. Abhandlung. Ann. Phys., 34, 446–463.

[45] Zagonel, L. F., Mazzucco, S., Tence, M., March, K., Bernard, R., Laslier, B., Jacopin, G., Tchernycheva, M., Rigutti, L., Julien, F. H., Songmuang, R., Kociak, M. (2011). Nano Lett., 11, 568–573.

[46] Zhou, W., Zhu, P., Qu, W. (2019). Basic properties of calcined underground ant nest materials and its influence on the compressive strength of concrete. Materials, 12, 1191.

Vikram Awate, Lokeshwar Patel, Rashmi Sharma, A. K. Beliya,
Ratnesh Tiwari, Vikas Dubey, Neha Dubey

Chapter 8
Thermoluminescence Glow Curve Analysis of Mn^{4+}-Doped Barium Yttrium Oxide Phosphor

Abstract: Synthesis and characterization of Mn^{4+}-activated BaY_2O_4 phosphor are reported in this chapter. Synthesized phosphor was analyzed by thermoluminescence (TL) glow curve for variable concentration of dopant and 5–10 min UV exposure time. TL glow curve shows well-resolved peak centered at 95 °C and intensity is maximum for 1 mol% of doing ion. Similarly, for 10 min UV exposure, the same pattern is followed by TL glow curve. The corresponding kinetic parameters are calculated by using peak shape method.

Keywords: UV exposure, doping effect, TL glow curve

8.1 Introduction

Thermoluminescence (TL) is a radiation-induced defect related phenomenon occurred in crystalline materials. Thermoluminescence studies of irradiated solids plays a major role to determine the electron spin resonance, fission track and cosmogenic nuclides analysis of certain minerals. Moreover, the traces of the integral effects of burial time, ambient temperature and in-site irradiation history of solids are also determined by the analysis of TL glow curve [1–16].

In this chapter, synthesis and characterization of phosphor by TL method are reported for variable doping ion concentrations as well as two different UV exposure time (5 and 10 min).

8.2 Experimental

The synthesis method opted for preparation of phosphor was solid-state reaction method; here, $MnCO_3$, $BaCO_3$ and Y_2O_3 reagents were used. The proper stoichiometric amount is weighed and thoroughly mixed. Some heat treatment was given for calcination (1,000 °C for 2 h) and sintering (1,300 °C for 4 h).

8.3 Results and Discussion

Figures 8.1 and 8.2 show the TL glow curve of Mn^{4+}-doped phosphors with variable concentration of doping ion and it shows the concentration quenching after 1 mol% of doping ions. TL glow peak centered at 95 °C shows the formation of shallower trapping in the Mn^{4+} doped BaY_2O_4 phosphor.

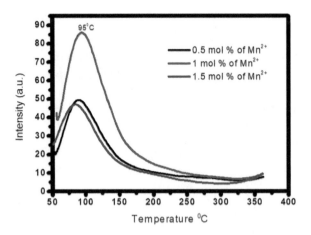

Figure 8.1: TL glow curve of Mn-doped BaY_2O_4 with the variation of Mn^{4+} concentration (5 min UV exposure).

Similarly, for 10 min UV exposure, same behaviors of TL glow curve were presented.

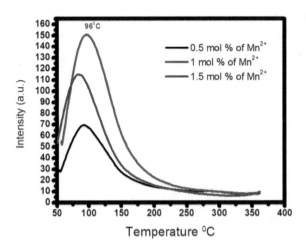

Figure 8.2: TL glow curve of Mn^{4+}-doped BaY_2O_4 with the variation of Mn^{2+} concentration (10 min UV exposure).

8.4 Conclusion

Here, rare-earth-free phosphor Mn^{4+} doped was prepared by the conventional method and it should be characterized by TL glow curve analysis with variable concentration of doping ions and 5–10 min UV exposure time shows a single broad peak which centered at 95 °C. It should be useful for environmental monitoring purpose with UV dose.

Keywords: environmental monitoring, Mn^{4+} doped, trapping

References

[1] McKeever, S. W. S. (1985). Thermoluminescence of Solids. Cambridge University Press, Cambridge.
[2] Chen, Y. Y., Li, S. H., Xie, X. N. (2005). Quartz luminescence geothermometer: preliminary studies on reconstruction of activities of thermal fluid flow in Yinggehai sedimentary basin, China. 11th International Conference on Luminescence and Electron Spin Resonance Dating, Cologne, Germany, 40–41.
[3] Guimon, R. K., Weeks, K. S., Keck, B. D., et al. (1984). Thermoluminescence as a palaeothermometer. Nature, 311, 363–365.
[4] Aitken, M. J. (1985). Thermoluminescence Dating. Academic Press, London.
[5] Adamiec, G., Aitken, M. (1998). Dose-rate conversion factors: update. Ancient TL, 16(2), 37_50.
[6] Sawakuchi, G. O., Okuno, E. (2004). Nucl. Instrum. Methods Phys. Res. B, 218, 217.
[7] Wintle, A. G., Murray, A. S. (2006). Radiat. Meas., 41, 369.
[8] Botter-Jensen, L., Mckeever, S. W. S., Wintle, A. G. (2003). Optically Stimulated Luminescence Dosimetry. Elsevier.
[9] Kaur, J., Suryanarayana, N. S., Dubey, V., Shrivas, R., Murthy, K. V. R., Dhoble, S. J. (2012). Recent Res. Sci. Technol., 4(8), 58–60.
[10] Dubey, V., Kaur, J., Dubey, N., Pandey, M. K., Suryanarayana, N. S., Murthy, K. V. R. (2017). Radiat. Eff. Defects Solids. doi: https://doi.org/10.1080/10420150.2017.1417410.
[11] Dubey, V., Kaur, J., Suryanarayana, N. S., Murthy, K. V. R. (2014). Res. Chem. Intermed., 40(2), 531–536.
[12] Dubey, V., Kaur, J., Suryanarayana, N. S., Murthy, K. V. R. (2013). Res. Chem. Intermed., 39(8), 3689–3697.
[13] Kaur, J., Suryanarayana, N. S., Dubey, V., Shrivas, R., Murthy, K. V. R., Dhoble, S. J. (2012). Recent Res. Sci. Technol., 4(8).
[14] Dubey, V. (2012). Thermoluminescence Study of Semaria Limestone of C.G.Basin. LAP LAMBERT Academic Publishing, ISBN 978-3-8473-4210-6.
[15] Dubey, V., Kaur, J., Suryanarayana, N. S., Murthy, K. V. R. (2012). J. Atoms Molecules, 2(4), 292.
[16] Kaur, J., Dubey, V., Suryanarayana, N. S., Murthy, K. V. R. (2012). Invertis J. Sci. Technol., 5(3), 162–166.

Rajani Indrakanti, V. Brahmaji Rao
Chapter 9
Theory of Luminescence and Materials

Abstract: Luminescence is the process to describe materials that absorb external energy and emit that energy to external surroundings. This can be observed with various excitation sources. The wavelength of emitted light can be used to study the characteristics of luminescent materials. This emission is defined as cold emission. It depends on the properties of a material. In this chapter, the basic concepts of luminescence are presented. The process can be understood with Stokes law and various mechanisms of luminescence. General characteristics of luminescence are discussed. The different luminescences are briefed with respect to different external sources and mode of excitation. They depend on the source of energy. These materials find numerous applications in electronics, electrical, computers, and biological industries while making optoelectronic devices, electronic, biological sensors and devices. Materials that follow luminescence properties are named as phosphors. The internal energy structure of various luminescent materials that are synthesized with rare earth elements, transition metals and semiconductors is discussed.

Keywords: luminescence, Stokes law, rare earth elements, transition metals, semiconductors

9.1 Introduction

We know that light is a form of electromagnetic energy obtained in different ways.

By supplying other forms of energy, one can prevail light energy, which can be obtained in two ways: incandescence and luminescence. Light obtained from thermal energy is called incandescence, that is, matter heated to a sufficient temperature will start to radiate. For example: (a) the electric stove's heater flame begins to glow in red hot, which is incandescence; (b) similarly when a filament is heated to a sufficiently higher temperature, then it glows brightly; (c) incandescence can be observed in the glowing of the Sun and stars.

Luminescence can be defined as an emission of light from a substance that leads from cold light. Therefore, it is defined as the cold-body radiation. It occurs due to chemical reactions and subatomic forces in crystals. These kinds of emissions are possible in substances that are in electrically excited state.

Acknowledgments: The author is grateful to the management of VNR Vignana Jyothi Institute of Engineering and Technology for their continuous support and encouragement.

https://doi.org/10.1515/9783110676457-009

9.2 Theory of Luminescence

"Luminescence" was introduced by Wiedemann (1888), which reflected an important property but not differentiated from reflected and refracted light. The duration of emission considerably increases the time period of oscillations [1–3]. In this process, the component of absorbed energy is released as an electromagnetic radiation. This emitted radiation can be in the visible or UV region of the electromagnetic spectrum. When an electromagnetic radiation interacts with matter, then the interaction follows mainly two steps: (1) the incident energy absorbed by the electrons of the material becomes excited to higher energy states; and (2) after their lifetime is over, the electrons were de-excited to their original state. During this time, the excess radiation will be released in the form of photons. Here, the time duration within the excitation and de-excitation may be short (10^{-8} s). Also de-excitation can occur within 10^{-8} s or via metastable state it can take more than 10^{-8} s. The duration of emission is subclassified into two processes: (1) fluorescence and (2) phosphorescence [4, 5].

Fluorescence is a spontaneous emission which takes place immediately as the life time reaches to 10^{-8} s. It depends on the intensity of absorption of radiation and independent of temperature.

Keywords: electromagnetic energy, phosphorescence, fluorescence

Materials having a shorter duration are called fluorescent materials. The phosphorescence process continues for some period of time even after the lifetime is completed.

If the time between absorption and emission is shorter, then distinguishing between fluorescence and phosphorescence will be difficult. Thus, phosphorescence can be observed in two ways: when time duration $\tau < 10^{-4}$ s and $\tau > 10^{-4}$ s. Therefore, materials having a longer duration are called phosphorescent materials. Thus, the theory of luminescence correlates the interaction of radiation with matter [4]. Here, light-emitting centers are defined as luminescence centers. Figure 9.1 shows the process of luminescence.

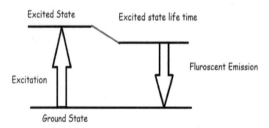

Figure 9.1: Luminescence.

In 1852, Stokes law or law of luminescence was formulated as follows: "The wavelength of the emitted light is greater than the exciting radiation" . According to Stokes law, the wavelength of emitted photons depends on characteristics of the substance and not on the incident wavelength [6]. The wavelength of emitted photons can be in the visible, ultraviolet (UV) and infrared spectrum region.

9.3 Features of Luminescence

Radiative or nonradiative transitions can be observed during the action of luminescence. In an efficient luminescent material, the radiative transitions prevail than the nonradiative transitions. On the other side, metastable states are created due to the addition of impurities so that traps are produced in the crystal. This causes delay in the luminescent emission as the electron–hole recombination takes place through traps.

Keywords: luminescence center, radiative transitions, metastable states

Largely, this type of recombination will be nonradiative. The mechanism of luminescence can be observed in few ways which are briefly discussed further.

9.3.1 Luminescence Based on Excited State Within the Molecules

The features of excited energy levels within molecules can be explained through conversion of energies among the states. When a molecule absorbs energy and gets excited, there are different ways to return to the ground state. The de-exciting of a molecule to the ground state can be understood by the Jablonski diagram. There are certain processes such as fluorescence, phosphorescence, internal conversion and intersystem crossing observed through this pictorial representation. For understanding, an example of Jablonski diagram is shown in Figure 9.2.

If the excess energy is liberated through photons between two states of same spin, then fluorescence is observed. And if it occurs between two states of different spins, then phosphorescence can be observed. These transitions are radiative.

Some nonradiative transitions are also possible. As the vibronic motions are present within a molecule (or atoms), multiple vibrational states develop within the state [7]. Due to this, singlet and triplet states are compiled and formed. This is clearly indicative as the order of energy state increases.

Keywords: Jablonski diagram, internal conversion, intersystem crossing, vibronic motions

The vibrational relaxation occurs in a short span of time. Here, the molecules transfer energy to other particles by means of collisions [8]. By collision, they return to ground state and transition is nonradiative [9–11]. Because of collision, there will be a change in bond length and angles of atoms within molecules.

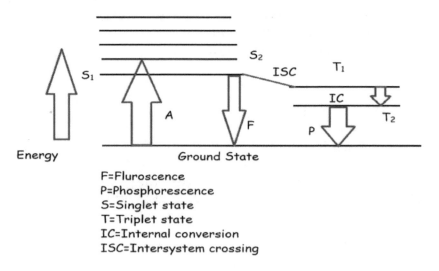

F=Fluroscence
P=Phosphorescence
S=Singlet state
T=Triplet state
IC=Internal conversion
ISC=Intersystem crossing

Figure 9.2: Jablonski diagram.

9.3.2 Luminescence Based on Configurational Coordinates

In general, the de-excitation of electron to the ground state by emitting a photon of frequency is equal to the difference of two energy levels, which is against Stokes law. In luminescent materials, parent crystal, luminescent ion, that is, activators are present. These activators are impurity atoms that are doped into the parent lattice. Counting on activators, emission and excitation time between energy levels will change so that features of the luminescent material are altered [12, 13]. Apart from activators, neighboring atomic vibrations affect the properties. This model explains by choosing a luminescent ion, nearest adjacent atomic sites, that is, luminescent center. The number of vibrational modes can be calculated by combining normal coordinates specifically known as configurational coordinates, Q. It is drawn with respect to the total energy of electron in a particular state with respect to configurational coordinates.

Keywords: activators, absorption and emission bands, configurational coordinate model

The amount of energy can be calculated from the sum of the electron energy and the ion energy. The model is represented in parabolic shape due to harmonic vibrations of ions; transitions are marked by vertical lines as shown in Figure 9.3. Stokes shift, temperature, width of absorption and emission bands can be computed. When the width of bands is wider correspondingly Stokes shift value will be large. In Figure 9.4, it is represented as Q_0.

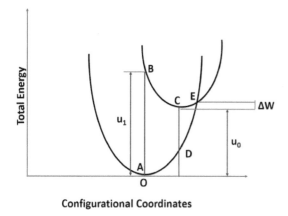

Figure 9.3: Configurational coordinate model.

9.3.3 Luminescence Based on Energy Band Formation in Solids

The luminescence of solid materials can be interpreted from band theory of solids. The electrons occupy certain discrete orbits based on the rules of quantum mechanics. They will have discrete energy levels. When a pulse of radiation in the form of photons is incident on a material, vacancies are created in occupied orbitals. Depending on the amount of photon energy, the electron will be shifted to higher energy level. Then an electron from an outer level will jump and occupy the vacant energy level. Once the lifetime of excited electron is over, the electron jumps to the ground state, and the difference in energy is radiated in the form of photon to the surroundings. In certain transitions, this emitted energy will be in the visible region.

The coulombic forces of attraction or repulsive forces of atoms start coming close together in solid. When the equilibrium state is reached, the atoms occupy regular sites and formation of solid takes place. During this process, the outermost orbitals of neighboring atoms start to overlap. As two electrons cannot occupy the same energy level, splitting of energy levels starts out inside the solid. These energy levels also split into sublevels [14, 15].

In a solid material, thousands and thousands of atoms are grouped together. The outermost orbital energy levels of neighboring atoms overlap with each other. Also they split into sublevels of slightly varying energies. Now an energy band is formed due to the outermost orbital of electrons. Either electrons are occupying or not occupying the energy levels, the band formation takes place. This band of energy is called valence band. There exists another band above the valence band known as conduction band. The difference of energy between the highest level of valence band and the lowest level of the conduction band is called energy band gap (E_g).

Depending upon the energy band gap between the valence band and conduction band, the solids are classified into (1) conductors, (2) semiconductors and (3) insulators.

Conductors: No forbidden gap is available between the two bands. Also the valence band overlaps with the conduction band in metals.

Keywords: forbidden energy gap, valence band, conduction band

Insulators: Large energy gap is observed in insulators. Here, the valence band is completely full and conduction band is completely empty.

Semiconductors: In case of semiconductors, the forbidden gap shown in Figure 9.4 between the valence band and conduction band is small compared to insulators and more as compared to conductors. At room temperature, some electrons jump from valence band into conduction band.

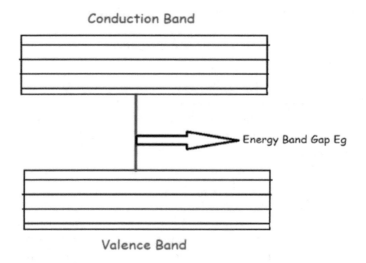

Figure 9.4: Splitting of energy levels and formation of valence band and conduction band.

Small amount of current flows in the crystal by applying electric field. In the above, luminescence can be observed in semiconductors when the electrons break covalent bonds and rise to conduction band after incident energy is absorbed. Then electron–hole pairs are created. This electron–hole recombination releases energy to the surroundings as radiative or nonradiative. These are discussed in Section 9.5.

9.4 Different Types of Luminescence

Several types of luminescence can be obtained with various incident radiations. These incident radiations can be photons, sound, electrical energy, mechanical energy, nuclear radiation and so on.

Keywords: semiconductors, electron–hole recombination, excitation

Depending upon the type of radiation, luminescence [16, 17] is differentiated based on the mode of excitation, and few of which are shortly discussed further.

9.4.1 Photoluminescence

Photoluminescence is observed if the light emission is observed due to the excitation of electromagnetic radiation photons. The emission continues for minutes or hours. Thus, it is also called as continual luminescence.

It has exclusive applications in optoelectronic devices, biological markers and so on at a large scale, road safety and exit marking.

9.4.2 Bioluminescence

The production of light due to a chemical reaction within a living organism is called bioluminescence. It is a case of chemiluminescence. In the process of a biochemical reaction, the light is produced. In the deep ocean layers where sunlight cannot reach, light is produced due to chemical reactions between living organisms, for example, fish, bacteria and jellies.

9.4.3 Sonoluminescence

If a gas bubble is driven in a liquid solution at ultrasonic frequencies which is acoustically suspended, then sonoluminescence can be observed. This results in bubble collapse, cavitation and light emission, and is used in shock temperature measurements of liquid hydrogen and to study shock-wave properties of solids.

9.4.4 Electroluminescence

When an electric field is passed through a material, it emits photons. These emitted photons exhibit both optical phenomena and electric phenomena. Usually, the devices are constructed by employing organic or inorganic materials. In general, semiconductors of wide band gap are in use which allow release of light.

Examples are ZnS, Mn, powdered zinc sulfide doped with copper, thin-film zinc sulfide, semiconductors containing group III–V elements.

9.4.5 Chemiluminescence

This is similar to bioluminescence, which occurs as a result of chemical reaction. In some cases, heat energy is also evolved. This can be observed in glow sticks, lightening objects, children's toys and analysis of inorganic species in liquid phase.

9.4.6 Triboluminescence

When a chemical bond is broken due to the application of force, then the separation or reunification of electric charges takes place inside the material.

Examples are diamond and quartz which begin to glow while being rubbed.

Keywords: bioluminescence, sonoluminescence, electroluminescence, chemiluminescence

This is one aspect of optical phenomenon in which light is rendered due to the breaking of chemical bonds.

9.4.7 Cathode Luminescence

In cathode luminescence, the electrons collide with a luminescent material such as a phosphor and emit the photons. Both optical and electromagnetic phenomena can be observed. If the wavelengths of emitted photons are in the visible region, then it can be envisioned; for example, the television screen made of phosphor generates light by electron beam scatterer using a cathode ray tube.

9.4.8 Radioluminescence

The bombardment of a material with ionizing radiation causes generation of light which is defined as radioluminescence. This can be observed in alpha rays, beta rays, gamma rays, nuclear reactors and radioisotopes. For example, the radioluminescent paint is used for clocks and dials that enable them to understand in the dark.

9.4.9 Fluorescence

When an electromagnetic radiation interacts with a substance, fluorescence can be observed when the lifetime of an atom reaches 10^{-8} s as discussed earlier. If the emitted light has larger wavelength than the absorbed radiation and it is in the visible region, then the fluorescent substance gets a discrete color. These materials cease to glow when the incident radiation is halted. Their applications can be observed in the mechanisms of geology, biology, medicine, chemistry, physics and material science.

9.4.10 Phosphorescence

The emission of light at a slower rate and lower intensity after absorbed radiation. This can be due to the forbidden energy levels in certain materials. They glow although light source is removed. Some of the examples are toys, stickers, paint, wristwatch and clock dials that glow in the dark.

9.4.11 Thermoluminescence

This is another form that can be observed in definite crystalline materials. In this type, the absorbed energy of electromagnetic radiation is released as light when a material was heated. This is also different from the blackbody radiation that is helpful in measuring heat parameters, particularly in ceramics.

9.5 Luminescence Materials

The substances that emit electromagnetic radiation in the UV, visible and infrared regions of the spectrum are called luminescent materials.

Keywords: thermoluminescence, radioluminescence, cathode luminescence

Phosphor is a material that can emit light in infrared to UV regions beneath the external stimulation. Under some conditions, this can be observed in visible region. The incident energy can be the form of high-energy electron, photons or electric field. Several forms of compounds like organic, inorganic and organometallic compounds are of high interest. There are several reports on these compounds according to the recent survey.
(a) **Organic compounds**: Aromatic hydrocarbons and derivatives, diphenylpolyenes, few amino acids and so on.

(b) **Inorganic compounds**: Lanthanide ions, Nd, Mn, Ce, Sn and so on, which are doped glasses, crystals like ZnS, GaS, GaP and so on.
(c) **Organometallic compounds**: Porphyrin metal, ruthenium and copper complexes with lanthanide ions.

All the abovementioned compounds exhibit features of luminescence. In this report, inorganic compounds are discussed.

9.5.1 Rare Earth Ions

Rare earth ions have gained research interest due to their usage as conversion materials in light-emitting diodes (LEDs) [18–20]. It can be observed that white LEDs are based on the partial conversion of the blue emission using one or more phosphor materials. Also violet color LEDs are employed as primary sources of light [21, 22]. This can efficiently absorb the excitation energy and pass it to the rare earth ion [23, 24]. When the central atom is an anionic group and oxygen atom and if rare earth ions substitutes the position of cations, then it is effective to transfer energy from anioninc to rare earth. This increases the luminous efficiency of rare earth ions.

Same parity is maintained by all states of a single 4f configuration so that optical transitions are forbidden. It can be due to transitions within the various energy levels are regulated by quantum mechanics. Here the symmetry and parity rules are followed by the crystal lattice. Accordingly, transitions can be strong or weak. The transitions are weak if lattice lacks inversion symmetry and of opposite parity. Magnetic dipole transitions are allowed among the similar parity through the selection rules. These are weaker in magnitude in comparison with electric dipole.

Keywords: organometallic compounds, rare earth ions

In odd-parity vibrations, electron–phonon interactions create the same configuration alloy in comparison to odd-parity mixing which is significant toward 3d ions [25]. The electronic transitions for rare earth ions constitute sharp lines.

9.5.2 Transition Metal Ions

In transition metal ions, the 3d series exhibit more substantial interaction with lattice environment when compared to 4f ion as no equivalent shielding by 5s, 5p outermost shells are present. The splitting of L, S multiplets are through static electric field. And smaller L–S coupling leads to weaker interaction. Here intra-transitions are weakly allowed by combining through odd-parity lattice fields. After parity, the transitions should follow the selection rule $\Delta S = 0$. Depending on the type of crystal-field interaction, zero phonon lines can be remarked when transitions are purely electronic. It has been reported that odd parity vibrations feature high transition probability because of configuration admixing [26].

9.5.3 Semiconductors

In inorganic semiconductors, recombination of electrons and holes gives rise to luminescence. The transitions from transition metal ions, and rare earth ions in semiconductors can be observed. Efficiency is observed in direct band gap semiconductors than indirect gap semiconductors. The transition probability requires conservation of wave vector (K) followed in direct band gap semiconductors. The existence of phonons plays a key role in indirect band gap semiconductors, which leads to nonconservation of wave vector (K). Luminescence can also be observed in excitons, donor-acceptor recombination, energy level structures which are shallow in nature. If electron affinities and band gap energies exist relevant to energy levels that come in band gap, then deep-level luminescence in semiconductors can be noticed [27]. Working of PN junction diodes, transistors and LED laser diodes is based on recombination. This has gained substantial importance in solid-state physics.

Keywords: transition metal ions, recombination, direct band gap, indirect band gap

9.6 Conclusion

This chapter discussed the concept of luminescence and Stokes law. Various types of luminescence with examples are given. The general characteristics of luminescence are presented. Luminescence in both inorganic materials such as rare earth ions, transition metal ions and semiconductors is discussed. The conditions for obtaining luminescence in rare earth ions, transition metal ions and semiconductors are also presented.

References

[1] Vij, D. R. (1998). Luminescence of Solids. Plenum Press, New York.
[2] Fouassier, C. (1984). Luminescence-Encyclopedia of Inorganic Chemistry. Academic Press, New York.
[3] Mckeever, S. W. S. (1985). Thermoluminescence of Solids. Cambridge University Press, Cambridge.
[4] Bube, R. H. (1951). The role of chlorine in the luminescence of ZnS: Cu phosphors. J. Chem. Phys., 19, 985.

[5] Ad Omar, M., Patterson, H. H. (1999). Luminescence Theory – Encyclopedia of Spectroscopy. Academic Press, New York.
[6] Marton, L. (ed.). (1959). Methods of Experimental Physics – Solid State of Physics. Academic Press, New York.
[7] Kitai, A. H. (1993). Solid State Luminescence. Chapman & Hall, London.
[8] Tsutsui, T. (1999). Fundamentals of luminescence. In: Shionoya, A. S., Yen, W. M. (Eds.). Phosphor Handbook. CRC Press, Boca Raton, FL, USA, 63–64.
[9] Murthy, K. V. R., Virk, H. S. (2014). Luminescence phenomena: an introduction. Defect Diffus. Forum, 347, 1–34.
[10] De, S., et.al. (2005). Electrooptical response of a single-crystal diamond ultraviolet photoconductor in transverse configuration. App. Phys. Lett., 86(21), 1–3.
[11] Collins, A. T., Cheng-Han, L. (2002). Misidentification of nitrogen–vacancy absorption in diamond. J. Phys. Condens. Matter, 14, L467–L471.
[12] Collins, A. T. (2000). Spectroscopy of defects and transition metals in diamond. Diam. Relat. Mater., 9, 417–423.
[13] Sastry, M. D., et.al. (2008). Non-linear optical properties of coloured diamonds: observation of frequency up-conversion. Diam. Relat. Mater., 17, 1288–1291.
[14] Babu, C. K., et.al. (2012). Synthesis and characterization of SrS:Eu,Ga phosphor. Int. J. Sci. Innov. Discov., 2(3), 231–235.
[15] Poisson, J. (2010). Raphaël Dubois, from pharmacy to bioluminescence. Rev. Hist. Pharm, 58 (365), 51–56. ISSN 0035-2349. PMID 20533808.
[16] Thomas, E., Williams, L., Original Types of Miners' Flame Safety Lamps, Welshminerslamps.com. 2013.
[17] Smiles, S. (1862). Lives of the Engineers. Volume III (George and Robert Stephenson). John Murray, London, 107. ISBN 0-7153-4281-9. (ISBN refers to the David & Charles reprint of 1968 with an introduction by L.T.C. Rolt).
[18] Joos, J. J., Meert, K. W., Parmentier, A. B., et al. (2012). Thermal quenching and luminescence lifetime of saturated green $Sr_{1-x}Eu_xGa_2S_4$ phosphors. Opt. Mater. (Amst), 34(11), 1902–1907.
[19] Li, Q., Liu, Z. P., Li, X. J., et al. (2016). Synthesis and luminescence properties of Sr2CeO4: Eu^{3+}, Tb^{3+} phosphors. Dig. J. Nanomater. Biostruct., 11(1), 313–319.
[20] Iftimie, S., Mallet, R., Merigeon, J., et al. (2015). On the structural, morphological and optical properties of ITO, ZnO, ZnO:Al and NiO thin films obtained by thermal oxidation. Dig. J. Nanomater. Biostruct., 10(1), 221–229.
[21] Qin, C., Huang, Y., Shi, L., et al. (2009). Thermal stability of luminescence of $NaCaPO_4$: Eu^{2+} phosphor for white-light-emitting diodes. J. Phys. D: Appl. Phys., 42(18), 185105.
[22] Chiu, Y. C., Liu, W. R., Chang, C. K., et al. (2010). $Ca_2PO_4Cl:Eu^{2+}$:an intense near-ultraviolet converting blue phosphor for white light-emitting diodes. J. Mater. Chem., 20(9), 1755–1758.
[23] Wu, B., Zhang, Q., Wang, H., et al. (2012). Thin copper oxide nanowires/carbon nanotubes interpenetrating networks for lithium ion batteries. Cryst. Eng. Comm., 14(6), 2087–2094.
[24] Okamoto, S., Yamamoto, H. (2003). Luminescence of rare-earth doped alkaline-earth titanates. J. Lumin., 102, 586–589.
[25] Shen, J., Sun, L. D., Yan, C. H. (2008). Luminescent rare earth nanomaterials. Dalton Trans., 42, 5687–5697.
[26] Avram, N. M., Brick, M. G. (2013). Optical Properties of 3d-Ions in Crystals. Springer, Berlin.
[27] Lu, W., Lustig, W. P., Li, J. (2019). Luminescent inorganic-organic hybrid semiconductor materials for energy-saving lighting applications. Energy Chem., 1(2), 100008.

Shalini Patil
Chapter 10
Luminescence: Phenomena, Applications and Materials

Abstract: Today, luminescence of nanoparticles, organic light-emitting diodes, imaging plates, phosphors for solid-state lighting, up-conversion luminescence, nonradioactive self-luminous paints, thermoluminescence dating, mechano-luminescent damage sensors and theoretical studies on luminescence are attracting attention from a large number of researchers worldwide. Luminescence is interdisciplinary in nature and it has great potential for basic and applied research.

Keywords: luminescence, incandescence, light emission

10.1 Introduction

Luminescence is an old field of scientific research. This is one research field where diverse application areas exist, which range from radiation monitoring for health and safety, phosphors for lamps and display purposes to X-ray imaging to other means of medical diagnostics.

Luminescence is defined as an emission of light which is in excess of that attributable to the blackbody radiation and persists considerably longer than the period of the optical radiation. There are two main processes which cause the emission of light. Incandescence means heating an object to such a high temperature so that the atoms become highly agitated leading to the glowing of the body. The light from the tungsten filament lamp or a burning piece of coal comes under this category. A cooler and more efficient mechanism of light emission is luminescence. In this mechanism, the light output per unit energy input is much larger than that in the case of incandescence. It is because of this that for the same degree of illumination, mercury tube lights consume much less electricity than the filament bulbs. Luminescence is a convenient term to cover both fluorescence and phosphorescence. It is the emission of visible light from a material when excited by X-rays, ultraviolet (UV) light, electron beam and heat of friction. Fluorescence is the emission of visible light by material under the stimulus of visible or invisible radiation of shorter wavelength. If the fluorescent glow persists for an appreciable time after the stimulating rays have been cut off, this afterglow is termed "phosphorescence." Sometimes, the phosphorescence may differ in color from the original fluorescence.

In 1652, Zechi made an important contribution to the understanding of photoluminescence. He observed that the color of the phosphorescence light in a material was independent of the color of the exciting light. About 200 years later, Stoke established fluorescence, the nature of luminescence, during excitation. He showed that the incident and emitted light differed in color.

10.1.1 Classification of Luminescence

There is a variety of luminescence phenomena observed in the nature or in man-made materials. The nomenclature given to these is invariably related to the exciting agent that produces the luminescence (Figure 10.1).

Keywords: fluorescence, phosphorescence, luminescence classification

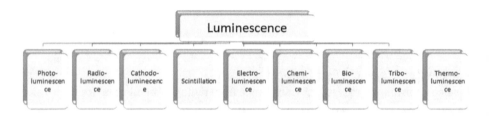

Figure 10.1: Flowchart of Different Types of Luminescence.

(i) Photoluminescence: This is the emission produced by excitation with the light photons. The fluorescent lamp used in household and general lighting is the principle example of this type. One of the high technologies called Light Amplification by Stimulated Emission of Radiation (LASER) is a kind of photoluminescence. About 254 nm UV radiation from mercury vapor discharge is absorbed by one of activator impurities in the phosphor coated on the inner side of the glass tube. By adjusting the relative concentration of these activator impurities, one can produce desired modification in the color of the light.

(ii) Radio-luminescence: When the excitation energy is provided by X-rays or nuclear radiations, the resulting luminescence is called radio-luminescence.

(iii) Cathodo-luminescence: When electron beams generated at the electrical cathodes are excited, the emission produced is called cathodo-luminescence. The screen of cathode ray tubes and television tubes glow by this kind of emission.

(iv) Scintillation: This phenomenon is same as radio-luminescence. It is called scintillation because it is used as a technique to detect individual light pulses generated by the incidence of each X-ray or gamma ray or a nuclear particle. Thallium-activated sodium iodide is a well-known scintillation detector used for gamma-ray spectrometry. The intensity of the scintillation is directly proportional to the energy of the incident gamma ray photon.

Keywords: injection luminescence, photoradioluminescence, cathodoluminescence

(v) Electro-luminescence: There is another type of electro-luminescence which is known as injection luminescence. In this type, electrons are injected from an external supply across a semiconductor p–n junction. On applying a DC voltage across the junction, such that the electrons flow to the p-region, luminescence is produced by the electron–hole recombination in that region. The light-emitting diodes (LED), which are now commonly used as display devices in many scientific instruments, are based on this principle.

(vi) Chemi-luminescence: Some chemical reactions are the source of luminescence. All chemical molecules are not capable of luminescence. Lyo-luminescence which is caused during the dissolution of certain compounds which have been bombarded by X-rays beforehand is a kind of chemi-luminescence. A well-known example of this type is X-irradiated NaCl, which emits a flash of light when quickly dissolved in water.

(vii) Bioluminescence: Biochemical reactions inside the cell of the living organism can produce electronic excited states of the biomolecules. Fireflies, glowworms, some bacteria and fungi and many sea creatures are striking examples of luminescence in living beings.

(viii) Triboluminescence: A large number of inorganic and organic materials, when subjected to mechanical stress, emit light. This phenomenon is called triboluminescence. It has also been named as mechano-luminescence.

(ix) Thermoluminescence: When the X-rays or gamma radiation or UV rays are exposed to a thermoluminescent material, the light produced by subsequent heating of material is called thermoluminescence.

Keywords: biomolecules, triboluminescence, chemiluminescence

10.1.2 Materials

There are many groups of luminescent materials: thermoluminescence dosimetry phosphors, phosphors for luminescent lighting, phosphors for X-ray intensifying screens, X-ray storage phosphors for digital radiography, broad band light emitting crystals for tunable lasers, luminescent glasses and glass ceramics for lasers, fiber optics, solar concentrators and lighting electro-luminescence thin films, cathode ray phosphors for projection TVs, scintillating materials, optical sensors for environmental monitoring, long-lasting phosphorescence materials for indication and decorations. In these materials, the role of impurities (activator) is important.

10.1.3 Phosphors

Phosphors are solid luminescent materials that emit photons when excited by an external energy source, such as an electron beam (cathodo-luminescence) or UV light (photoluminescence). Phosphors are composed of an inert host lattice which is transparent to the excitation radiation and an activator. The process of luminescence occurs by absorption of energy at the activator site, relaxation and subsequent emission of a photon and a return to the ground state. The efficiency of a phosphor depends on the amount of relaxation that occurs during the activation and emission. Relaxation is the process in which energy is lost to the lattice as heat; it needs to be minimized in order to extract the highest luminous efficiency. The luminous efficiency is defined as the ratio of the energy emitted to the energy absorbed.

Keywords: types of materials, phosphors, energy band

10.1.4 Properties of Phosphors

Most phosphors consist of a host composition plus the activator, added in carefully controlled quantities. The activator is a substitutional defect. Therefore, it is essential that the charge on the substitutional cation is equal to that of the host lattice cation.

We denote a phosphor as $M_a YO_b : N_x$
where M is the cation,
YO_b is the anion
and N is the activator.

It is known that the proper choice of host and activator is essential to obtain an efficient phosphor. Consider the case where 100 photons are incident upon the phosphor. Out of them, few are reflected, some are transmitted and if the phosphor is an efficient combination of host and activator, most of the quanta are absorbed. Some become deactivated via relaxation process. The quantum efficiency is defined as:

$$QE = \text{Photons emitted/photons absorbed}.$$

It is easier to measure intensity of photons emitted as a function of wavelength. This gives:

$$QE = (Id\lambda) \text{ emission}/(Id\lambda) \text{ absorption}.$$

Phosphors, which have 80% or greater efficiency, are considered to be efficient phosphors.

Keywords: UV light, absorption energy, activators

It is known that new phosphors are essential because of technological progress. Also a good phosphor for electronic or UV excitation is not necessarily a good choice for excitation in vacuum UV. Until now, to produce light with a fluorescent lamp, it is necessary to put mercury inside the lamp to generate UV photons at $\lambda = 254$ nm in order to excite the phosphor coated lamp inner surface. But in a near future, it will be very harmful for the environment. It should be replaced with a mixing of rare gases (xenon and neon) which emit vacuum UV photons from 147 to 190 nm. Luminous efficiency, color rendering and longevity are all properties which depend on the nature and the quality of phosphors.

10.1.5 Applications

There is a worldwide trend in the physical sciences toward applied research, particularly relevant to environmental and energy conservation problems. The possible applications depend on improvements in efficiencies and stabilities of inorganic luminescent materials.

Luminescent materials popularly known as phosphors have found their applications in various fields. Phosphors for lamps, color TV screens, long-lasting devices, laser host, scintillators and pigments luminescent paints and inks, nonlinear optical materials, CRO (Cathode-Ray Oscilloscope), LED, chemical analysis, watch dial, standard light source, X-ray imaging, personal monitoring and opto-electronic devices.

Keywords: fluorescent lamp, efficient phosphor, UV photons, long-lasting devices, nonlinear optical materials.

References

[1] Harvey, E. N. (1957). A History of Luminescence. American Philosophical Society.
[2] Shionoya, S., Yen, W. M. (1998). Phosphor Handbook; Phosphor Research Society. CRC Press, Boca Raton, FL.

Vinod Kumar Verma, K. K. Dubey
Chapter 11
Thermoluminescence and Spectral Studies of Some Geological Crystals

Abstract: This chapter reports the thermoluminescence (TL) and spectral studies of geological crystals such as rock salt, quartz, sphalerite, carborundum, fluorite and calcite crystals. The peak in the TL glow curve of rock salt crystals is at two temperatures, 177 and 196 °C. The peaks in the TL glow curve of quartz crystals occur at 220 and 370 °C. The peaks in TL glow curve of sphalerite crystals occur at 86 and 210 °C. The peaks in TL glow curve of carborundum crystals are found to be at 98 and 250 °C. A single peak in the TL glow curve of fluorite crystal occurs at 290 °C. The peak in the TL glow curve of calcite crystals are at 110, 260 and 375 °C. The peaks in the TL emission spectra occur at 440, 460 and 550 nm for quartz, rock salt and carborundum crystals, respectively. The TL emission spectra of fluorite crystals have two peaks, one at 475 nm and the other at 575 nm. The peak in the TL emission spectra of sphalerite and calcite crystal has only one peak lying at 520 and 625 nm, respectively.

Keywords: thermoluminescence, geological crystals, γ-radiation (^{60}Co) source

11.1 Introduction

Thermoluminescence (TL) is a common and widespread phenomenon of the two-third are known to exhibit TL (1.5). TL has been first mentioned be Robert Boyles in Royal Society of London in 1663. He described TL as "glimmering light" which is observed from heating diamond in the dark (by holding it against his) "nacked body." Weidemann and Schmidt [1–4, 12] were probably the first to carry out an extensive, controlled investigation of TL in a wide variety of natural and synthetic materials. An article was published which some of the possibilities of applied TL were suggested, including uses in geological stratigraphy, geological age determination and radiation dosimetry [5, 11]. The principal assumption is that the TL is a measure of the absorbed radiation close since the specimen was last healed, which in the case of a rock is its time of formation.

However, age determination is not the only utilization of TL in geology. In some instances, TL is more sensitive for detecting traces of radioactivity than more conventional means, for example a Geiger counter or a scintillation counter, some minerals are found to exhibit a particular glow curve shape when extracted from one area but the same mineral gives an entirely different glow curve it extracted from another.

Thus, TL has also found use in source identification. This chapter reports the TL of certain geological crystals like rock salt, quartz, sphalerite, carborundum, fluorite and calcite.

Keywords: radiation dosimetry, scintillation, quartz

11.2 Experimental

For the present investigation, the {geological crystals} were collected from different sources. Rock salt from Pakistan, quartz from Jabalpur (Gour River), sphalerite from Zawar, carborundum (SiC) from Jabalpur, fluorite from Chandidongri (Madhya Pradesh) and calcite ($CaCo_3$) form Gujarat. The crystals were reduced to a suitable size of $2 \times 2 \times 2$ mm^3 by grinding and polishing. The crystals were reduced to a suitable size and irradiated to gamma rays. TL glow curve were recorded with the help of routine setup consisting of an arrangement for heating the samples, light detector, signal processor and signal recorder. To achieve a more conclusive picture of the mechanism of TL spectra of these geological crystals, they are also recorded by TL emission spectra recorder.

11.3 Result

11.4 Discussion

For the present investigation, the geological crystals were collected from different sources. Rock salt from Pakistan, quartz from Jabalpur (near Gour River), sphalerite from Zawar (Rajasthan), carborundum from Jabalpur, fluorite from Chandidongri (Madhya Pradesh) and calcite form Gujarat.

11.4.1 Rock Salt

Natural rock salt shows a variety colors ranging from gray to black, red, brown, yellow, green, blue and violet. TL of the natural blue salt was observed by Kraatz-Koschlau and Wholer in 1889. The TL of impurity-doped NaCl crystal consists of a principal glow peak at lower temperature. It is suggested that the dipoles in the dislocation free region are responsible for this low-temperature glow peak. The higher-temperature glow peak consists of a combination of impurity vacancy dipoles with a negative ion vacancy, the complex being situated in the dislocation region.

Keywords: natural rock, rock salt, mechanism of TL

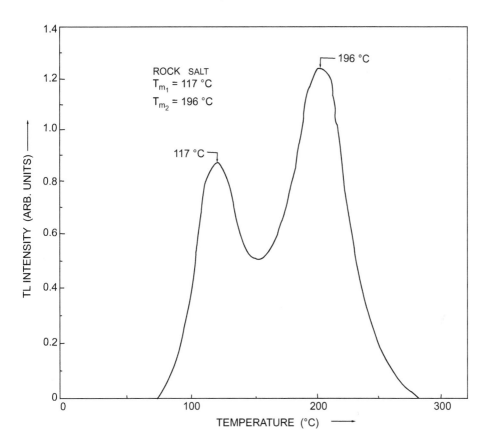

Figure 11.1: TL glow curve of rock salt crystal; it is shown that peaks in the TL glow curve occur at 177 and 196 °C.

11.4.2 Quartz

Quartz occurs in colorless, yellow, gray, brown to black, violet and pink forms, and these forms are called ordinary quartz. The color of smoky quartz, bleached by heating is again reproduced by irradiation. Colorless quartz turns into smoky quartz with radium irradiation. All the criteria of radiation coloration is not due to the effect of the radiation on the quartz itself, but on an inclusion of sodium silicate.

11.4.3 Carborundum

In the case of SiC crystal least, studies have been made on the TL. In case of TL, the emission has been found around 540 nm. This fact indicates that similar luminescence center is responsible for TL.

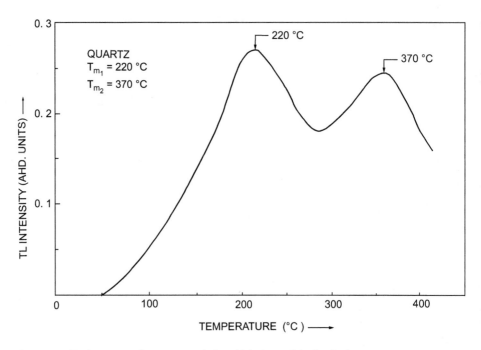

Figure 11.2: TL glow curve of quartz crystals for which the peak in the TL glow curve occurs at 220 and 370 °C.

11.4.4 Sphalerite

The TL spectrum of sphalerite phosphor has broad peak lying at 520 nm. It indicates that in the process of TL, the recombination of electrons take place with the holes trapped at acceptor levels of higher energy. In the TL in sphalerite crystals, the luminescence center is formed by zinc ions occupying different sites in the lattice and being perturbed by neighboring impurities or native defects [6–9].

11.4.5 Fluorite

The TL emission spectra of this material involving RE^{3+} emission has been explained during exposure to ionizing radiations, some of the released electrons are captured by the impurity RE^{3+}. The holes produced during the irradiation are in turn captured by the lost related defect centers. Some holes are also captured in impurity centers. The change conversion process of the centers involved in the emission can be written in the following way:

$$RE^{3+} + \bar{e} \rightarrow RE^{3+}$$

Chapter 11 Thermoluminescence and Spectral Studies of Some Geological Crystals — 237

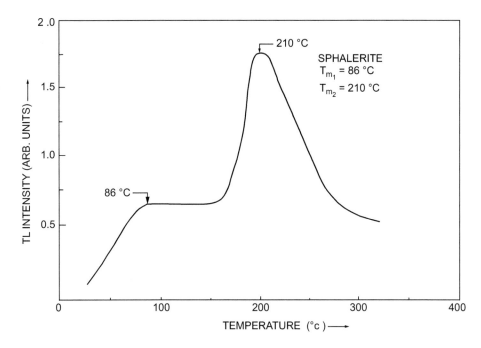

Figure 11.3: TL glow curve sphalerite crystals; it is shown that the peaks in the TL glow curve lie at 86 and 210 °C.

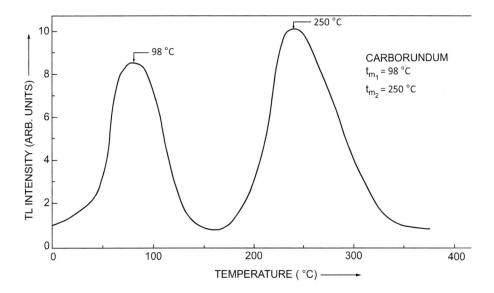

Figure 11.4: TL glow curve carborundum for these crystals; the peaks in the TL glow curve are found to be at 98 and 250 °C.

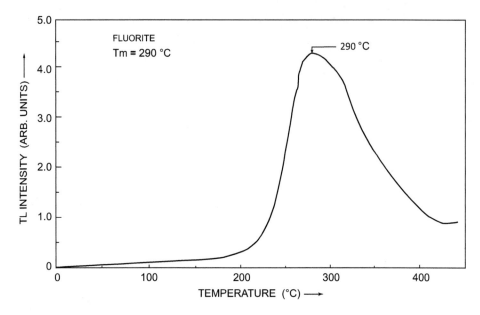

Figure 11.5: There is only one peak in the TL glow curve of fluorite crystals and they lie at 290 °C.

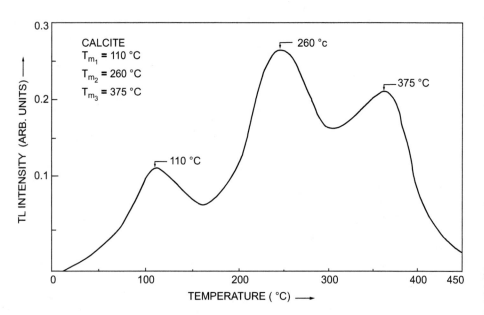

Figure 11.6: TL glow curve observed in the calcite crystals. It is seen that there are three peaks in the TL glow curve and they lie at 110, 260 and 375 °C.

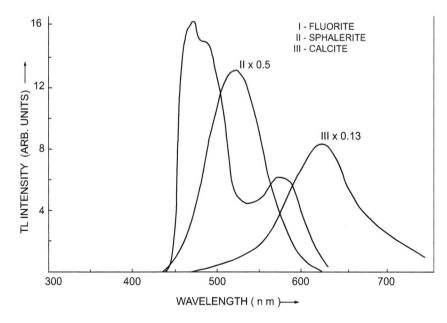

Figure 11.7: TL spectra of quartz, rock salt and carborundum crystals the peak of the emission spectra occurs at 440, 460 and 550 nm for quartz, rock salt and carborundum, respectively.

$$F + \text{hole} \rightarrow V_k$$
$$V_k \rightarrow F + F + \text{hole}$$

During heating

$$RE^{2+} + \text{hole} \rightarrow RE^{*3+} \rightarrow RE^{3+} + TL$$

where RE^{*3+} means RE^{3+} in an excited state.

11.4.6 Calcite

In calcite containing manganese, the manganese acts as the major luminescent center and the crystal glows in the orange region of the spectrum. Early measurements by Medin [10] showed some structure of the TL emission spectrum and also changes of the X-ray excited luminescence with temperature.

11.5 Conclusion

For a given radiations dose, the TL intensity is found to be different geological crystals. In the descending order of TL intensity, the crystals may be arranged as carborundum, fluorite, sphalerite, calcite, rock salt and quartz crystals. A preliminary investigation indicator that upto a given radiation dose the TL intensity increases with the radiation dose. The first peak is assumed to be related to the crystallinity of the specimen. It is, therefore, suggested that the dipoles in the dislocation free region are responsible for the low temperature glow peak. The higher glow peak consists of a combination of impurity vacancy dipoles with a negative ion vacancy, the complex being situated in the dislocation region.

Keywords: luminescent center, crystallinity of the specimen, thermoluminescence emission

References

[1] Aitkesh, M. J., Thompson, J., Flremings, J. (1968). Proc. 2nd Int. Cong. Lumi. Dosimetry U.S.A., C.C. NITS spring field, 364.
[2] Andirovitch, E. I. (1956). J.Phys. Radiat., 17, 705.
[3] Atari, N. A. (1982). Phys. Lett., A-90, 93.
[4] Cornu, F. (1908). Blue rock salt, Menues Jb. Miner. Geol. Palanot., 32.
[5] Danials, F., Boyd., C. A., Saunders, D. F. (1953). Science, 117, 343(54), 604 (in German).
[6] Desmukh, B. T., Moharil, S. V. (1985). Bull. Mater. Sci., 7(5), 447–457.
[7] Diehl, H., Grasser, R., Scharmann, A. (1968). Physikalisches Institute University Giessen. Germany.
[8] Frndel, C. (1945). Effector radiation on the elasticity of quartz. Amer, min, 30, 432.
[9] Garlick, C. F. J. (1949). Luminescence Material. Cberendon Press, Oxford.
[10] Medin, W. L. (1959). J. Chem. Phys., 32, 943.
[11] Randall, J. T., Wikins, M. H. F. (1945). Proc. R. Soc. London Ser. A. 184, 366, Trans fareday Soc, 35, 2.
[12] Widemann, E., Schmidt, C. C. (1995). Ann. Phys. Chem. neue folge.

Index

3D system structure 55
3-methacryloxy-propyl-trimethoxysilane 17

absolute temperature 49
absorption 109, 112, 116–117, 125
activator 230
Al_2O_3 150
aluminum 114
amino clay (AC) 25
amorphous 109, 111–112, 114, 125
amounts 109, 111
amplitude 117
angular 174
antenna effect 3
applications 172–173, 179
arch 134
area 173–174, 177–178
atoms 110–111, 227

background 158
band gap 112, 117, 125, 171, 176
basal state 119
beam 174–175
beta-diketone 11
between 172–174, 176
BFPP-Si 9
bifunctional rare earth 58
bioluminescence 197, 221, 229
biosensors 33
bipyramids 111
bis-silylated-bipy 10
bond 111
bottom-up 6
broad 173, 177–178

calculations 178–179
calix[4]arenes 9
calixarene 18
$CaMoO_4$ 153, 157
candidates 178
carbon 149
carbon dots (CDs) 28
carboxylate groups 65
carboxylate units 60
carboxylates 9
cathodoluminescence 194, 230

cation 110
CdS 140
CdS/chitosan 141
CdS/dendrimer 141
CdS/DSS 141
CdS/PPH 142
CDs@MOF-253 28
CdSe/ZnS 143
centered 176
chemiluminescence 195
chromaticity 109, 122, 125
CIE 57, 165
CIE coordinate 167
clearly 176–177
coated 149
co-condensation 17
color 109–110, 123, 125, 173
color rendering 231
Commission International d' Eclairage 122
composition 109–115
compound 172
computer 175
concentrations 110–112, 114, 119, 122, 124, 176
conduction 173
conduction band 219
connected 110
coordinates 122–123, 125
copper 175
co-precipitation 202
corrosion 114
coupled 175
covalent 119
criminal individualization 133
criteria of radiation 235
crystal field 2
crystal structure 60
crystalline 175–176, 178
crystalline phase 111
crystallinity of the specimen 240
cubic 175
Czochralski 175–176, 179

decades 172–173
decay 109, 124–126
deep-level luminescence 225
defects 116, 122, 125

degree 172
density 111, 116, 125
detect 174
detection 144
development 139, 157, 159
devices 172, 178–179
Dieke's diagram 2
diffusivity 173
dioxide 146
dispersed 178
display devices 110, 123
diurasil 17
doped 109–111, 114, 116–123, 125
dots 143, 149
double 179
down-conversion 189
dusting 159

earth 150, 153
edge 117
effect 173
effective energy 54
efficiencies 178
efficient energy transference 48
electric dipole 2
electric dipole–quadrupole 3
electrical 172
electrochemical sensors 33
electrochemiluminescence (ECL) 32
electrode 32
electroluminescence 195
electro-luminescence 229
electroluminescent 33
electromagnetic 174
electron spin resonance 211
electronic 172
electron–hole recombination 220
elements 150
emission 66, 110, 112, 119, 122–125, 137, 175–178
emission intensities 49
emission intensity 49, 54, 63
emission spectra 57, 236
emissive 4
energy 173–174, 177
energy gap 4
energy level 68
energy transfer 4
energy transference 66

energy transference method 48
enrichment 119
epitaxial 178
Eu^{3+} 17
Eu^{3+}@ UiO-bpy 31
Eu^{3+}@MIL-124 26
europium 109–112, 116–117, 119, 123, 125, 153
excited state 110, 239
excitons 225
exhibiting single emission 58
exponential 172
extreme temperature 51

faces 175
films 172
fingerprint 139, 143–144, 151, 157, 159
fingerprinting 158
fingerprints 143, 146, 153–154
fission track 211
fluorescence 119, 140, 147, 152, 193, 223
fluorescent lamp 231
fluorescent nanomaterials 132, 135
fluorescent probe 27
fluoride 28
forensic 139
forensic science 131
formation 174–175, 178
frequency 174
function 174–175
functionalization 22

gadolinium 5
gaps 173
Gaussian 176
geometry 174
germanium 173
Gibbs triangle 113
glass former 110
glasses 109–126
graphene 172, 177
graphics 174
great 172–174, 178–179
green 110, 113, 125, 176
ground 175

heat treatment 109, 112, 114, 122, 125
heavy metal 111
hendecahedral geometry 60
heterostructures 142

high refractive index 110–111
hole 174
HOMO 186
hybrid 12
hydrothermal reaction 58
hypersensitive 119

identification 139
illuminant 123
immunosensors 33
impressions 146
impurities 173, 178
incandescence 215
incidence 174, 176
increasing 176
individual 173–174, 176
industrial 175
instruments 174
intensities 176
intensity 109, 114, 116, 119, 122, 124–126
interesting 173–174
interstitial 111
investigation 173, 175, 178
ions 109–111, 116, 119, 121, 125
IR analysis 60
irradiated 172
ITO 20

Jablonski diagram 217

know 179

lamp 174–175
lanthanide 1, 66
lanthanide hybrids 22
lanthanide ion 47
lanthanide photofunctionalized 26
lasers 109, 112
latent 143–144, 147, 153, 157, 159
latent fingerprints 133
latent prints 142
law 173
lenses 174
levels 172
lifetime 124
ligands 5, 119
light 109–110, 113, 123
liquid 152
Ln^{3+} 25

Ln-MOF temperature sensor 52
LnMOFs 31
loop 134
lower doping 167
luminescence 14, 135, 157, 172, 176–177, 179, 215, 227
luminescence intensity 68
luminescence properties 49
luminescent 109, 137
luminescent behavior 54
luminescent center 218, 239
luminescent intensity 27
luminescent materials 223
luminescent MOFs 28
luminescent solar concentrators 23
luminophore 32
luminous efficiency 230–231
luminous temperature 46
luminous transitions 48
LUMO 186

magnetic dipole 1, 119
magnification 176
manufacturing 172
materials 172–176, 178–179
matrix 110–113, 115–117, 119, 122, 125, 176, 178
MCM-41 12
mechanical 5
mechanism 178
mechanism of thermoluminescence 234
mechanoluminescence 198
melting 110–111, 114
melt-quenching 109, 112
mention 173–174
mesoporous 12
metal organic frameworks (MOF) 26
metal organic frameworks 45
metal sulfide nanoparticles 20
metastable states 217
methacrylic acid 17
microporous 12
MIL-100 26
MIL-124 26
mm 175
Mn^{4+} doped 212
modifier 125
modifiers 110–111
MOFs 52, 55
molecular 109, 111

monochromator 174–175
morphology 10
multifunctional luminous 46
multiplying 177
multiwalled carbon nanotubes 22

nanocrystals 141
nanoforensics 132
nanohybrids 17
nanomaterials 20, 137, 153
nanoparticles 144, 149–150, 154, 175–176
nanophosphors 159
nanostructures 157
nanozeolite L 31
natural rock 234
NaYF$_4$ 152
network 110–112, 115
new 173, 178–179
nondestructive 172, 175
nonporous 147
nonradiative decay 49
numerically 177

observation 176
opaque 111, 113, 117, 125
optical 173, 178
optical properties 111
optics 174
optoelectronic 172
organic antenna 47
organic framework 52
organosilane 25
organosilicas 12
ORMOSILs 6
overlapped 115
oxidation 172, 176
oxidation states 110
oxide 149
oxygens 110

parameters 179
patent fingerprints 133
patterns of fingerprints 134
peaks 109–110, 115–117, 119, 125, 175–176
periodic 173
periodic mesoporous organosilicas 14
phase 109, 111–112, 115, 119, 122
phenomena 172–173, 178

phosphor 137, 151, 157, 211, 230
phosphorescence 5, 194, 216, 227
phosphors 110, 136
phosphorus 123, 125
photoluminescence 18, 136, 171–176, 178, 191, 206
photons 228
photophysical 1
photovoltaic 172
photovoltaic cells 205
photovoltaic effect 205
photovoltaics 25
physical 173, 178
physiological range 58
physiological temperature 69
PL 173–178
PL emission spectra 167
PL spectra 52
place 174
Pluronic P123 15
polarizability 111
polarizable 119
polyhedra 111
polymer 14
polymer light emitting diodes (LEDs) 33
polysilsesquioxane 19
polyurethane foams (PUFs) 27
population 111
porous 172, 176
position 174, 176
possible 173–174, 178
powders 158
power 174
print 157
prisms 174
process 173–176
properties 172–173, 178–179
purity 109, 123, 125

quantum 143, 178
quantum dots 138
quantum efficiency 33
quantum yield 9
quartz 234

radiation dosimetry 233
radiation monitoring 227
radiative 172
radioluminescence 197–198, 222

radius 109–110, 119, 121, 125
rare 150, 153
rare earth 109, 110
ratiometric temperature 46
ratiometric thermometer 66
ratiometric thermometers 45
recombination 236
red 109–110, 119, 123, 125
reddish-orange 110
region 110–111, 116–117, 122–123
relaxation process 230
research 172–173, 178
response 174
results 176
reticulum 109, 119, 121
review 173, 179
rock salt 239

samples 109, 112, 114, 124–125, 174–176, 179
Scherrer's formula 115
scintillation 228, 233
selectivity 158
semiconductor 20, 137, 139, 172–175, 178–179
semiconductor nanoparticles 133
semiconductors 125
sensing 27
sensitivity 145
sensitizer 57
sensor 28
shallower trapping 212
shows 174, 176
signal 174, 176
silica 18
silicate 110
silicon 171, 173
silicon oxide 140
silver 139
single-walled nanotube (SWNT) 30
SiO_2 144
size 173, 176
solid 173, 177
solid-state method 166
solubility 173
solvothermal 199
sol–gel 17, 203
sonoluminescence 197
source 174
spatial positions 51

spectra 109, 111–112, 116–117, 119, 122, 174, 176, 178
spectral shift 49
spectroscopic 46
spintronics 173
sputtering 172
start 174
states 176
stimulated 174
Stöber method 21
stoichiometric calculations 166
Stokes law 184, 216
Stokes shift 218
structural 109–112, 125
structure 58, 174–176, 178
studies 173, 178–179
support 178–179
surface 173, 176
symmetry 110, 119
symmetry and parity rules 224
synthesis 211
system 174

Tauc method 112, 117
Tb^{3+}@Cd-MOF 27
techniques 172, 176
technological 172
tellurite 109–110
tellurium 111–112, 124
temperature 68–69, 110–112, 119, 123, 125, 172, 175
temperature detectors 47
temperature increases 65
temperature outline 51
temperature sensors 45, 50, 65
temperature-dependent luminescence 48
temperature-dependent PL 52
terbium 153, 157
TESPIC 6
theoretical 179
thermal 5
thermal sensitivity 50
thermally determined elimination 54
thermoluminescence (TL) 211
thermoluminescence emission 239
thermometers 45
through 172–175, 177
TiO_2 146
titania 19

titanium 146
titanium dioxide 140
TL spectrum 236
total 177–178
transfer 173
transfer sensitization 52
transformation 125
transitions 109–110, 116, 119
transparent 111, 113, 125
triboluminescence 198
trifluoroacetylacetone 19
trigonal pyramids 111
tube 177
two 174, 176–178
type 174–175, 178

Uio-66(Zr)-(COOH)$_2$ 27
ultraviolet light 230
UV 176
UV-LED 25

valence 173, 176
valence band 219
variation 173, 176
vibrational 217

visualize 147
vital 174
vitreous 109–112, 119, 121, 125
volatile organic compounds (VOCs) 31
voltage 172

watts 174
wavelength 109, 119, 122–123, 174, 176
weight 111
whorl 134

xerogels 19
X-ray 173, 175–176
X-ray imaging 227, 231

YSBO phosphor 166
YSBO:Eu^{3+}-doped phosphor 165

Zachariansen 110
zinc 149
zinc sulfide 140
ZnO 147
ZnS 143